SpringerBriefs in Environmental Science

SpringerBriefs in Environmental Science present concise summaries of cutting-edge research and practical applications across a wide spectrum of environmental fields, with fast turnaround time to publication. Featuring compact volumes of 50 to 125 pages, the series covers a range of content from professional to academic. Monographs of new material are considered for the SpringerBriefs in Environmental Science series.

Typical topics might include: a timely report of state-of-the-art analytical techniques, a bridge between new research results, as published in journal articles and a contextual literature review, a snapshot of a hot or emerging topic, an in-depth case study or technical example, a presentation of core concepts that students must understand in order to make independent contributions, best practices or protocols to be followed, a series of short case studies/debates highlighting a specific angle.

SpringerBriefs in Environmental Science allow authors to present their ideas and readers to absorb them with minimal time investment. Both solicited and unsolicited manuscripts are considered for publication.

More information about this series at http://www.springer.com/series/8868

Mary J. Thornbush

The Ecological Footprint as a Sustainability Metric

Implications for Sustainability

Springer

Mary J. Thornbush
Faculty of Environmental Studies
York University
Toronto, ON, Canada

ISSN 2191-5547 ISSN 2191-5555 (electronic)
SpringerBriefs in Environmental Science
ISBN 978-3-030-62665-5 ISBN 978-3-030-62666-2 (eBook)
https://doi.org/10.1007/978-3-030-62666-2

This Springer imprint is published by the registered company Springer Nature Switzerland AG
The registered company address is: Gewerbestrasse 11, 6330 Cham, Switzerland

Preface

The Ecological Footprint is a popular sustainability metric that informs human consumption and can be set against bioproductivity at various scales as a land-based composite indicator. It was introduced by Rees and Wackernagel and, since the early 1990s, has been developed by members of the Global Footprint Network. When applied in evidence-based decision-making, the National Footprint Accounts at the national scale can provide information for policymakers and governments to establish regulations that ensure sustainability. It is, therefore, crucial that the Ecological Footprint and biocapacity accounting be recognised as a measure of human resource consumption (demand) set against natural capital (supply). Governments would do well to adopt it alongside financial accounts to represent natural capital and its use. Although, as a sustainability metric, the Ecological Footprint is considered to wholly represent only the environmental dimension as a biophysical indicator, arguably it can do more than that as a consumption-based indicator. Based on a time series since 1961, it is possible to track cross-temporal changes of land-type categories (crop land, grazing land, forest land, fishing grounds, built-up land, and carbon) of the Ecological Footprint and biocapacity accounting for a specific annual edition of the accounts. This information conveys whether a country is in ecological deficit or reserve and what may be contributing to such a national trend. Although the analysis is often executed at national scale, it is possible to access other scales. The study area in the Alexander Skutch Biological Corridor ('the corridor') of Costa Rica represents a local–regional case that is compared to its national environmental performance based on the Ecological Footprint and biocapacity. Farmers in the core corridor were surveyed in the case study using the Footprint Calculator as a measure of environmental performance on the consumption categories of food, housing, and transportation included in this global Footprint Calculator to address 15 main questions involved in the assessment. Thus, their resulting Footprints are discussed here through comparison to the national level as a special case in the corridor. By doing so, it is possible to discern the difference that regional Footprints can have from national figures. This becomes important especially for large countries that have local–regional differences in their consumption and production of natural capital.

Toronto, ON, Canada Mary J. Thornbush

Contents

1 Introduction 1
1.1 Overview 2
1.2 Theoretical Framework 2
1.3 Conceptual Framework 5
1.3.1 Equations 7
1.3.2 Data Sources 10
1.4 Quality Analysis 10
1.4.1 Criticisms 11
1.4.2 (Spatial) Scale 15
1.4.3 Aggregation 15
1.4.4 Data Quality 16
1.5 Brief Aims 18
1.6 Conclusions 20
Appendix 1.1: Footprint Calculator Survey (in English) 21
References 23

2 The Ecological Footprint 31
2.1 Ecological Footprint Evolution 32
2.2 Sustainability Metrics 35
2.3 Systems Approach 36
2.4 Integrated Sustainability 37
2.5 Conclusions 38
References 38

3 Biocapacity Accounting 41
3.1 Biocapacity as 'Carrying Capacity' 42
3.2 Framing Production 45
3.3 Ecological Overshoot 46
3.4 Conclusions 48
References 48

4 Case Study – Methods 51
4.1 Research Question 54
4.2 Case Aim and Objectives 55
4.3 Study Area Sampling 56
4.4 Data Collection 58
4.4.1 (1) Informed Consent 59
4.4.2 (2) Demographics 60
4.4.3 (3) Farmers' Surveys 60
4.4.4 (4) Follow-up Interviews 61
4.4.5 (5) Wrap-Up 62
4.5 Data Analysis 62
Appendix 4.1: Fieldwork Datasheet 63
References 65

5 Case Study – Findings 67
5.1 Findings with Discussion 68
5.2 Demographics 69
5.3 Farmers' Surveys 71
5.4 Follow-up Interviews 74
5.4.1 Q1: What Type of Crops Do You Grow, and in What Proportion? 74
5.4.2 Q2: How Much Do You Grow in a Year? 75
5.4.3 Q3: Where Do You Sell Your Crops? 76
5.4.4 Q4: Are You Participating in Fair Trade or Organic Farming? 77
5.4.5 Q5: Do You Use Fertilisers or Pesticides? 78
5.4.6 Q6: Do You Benefit from Payments for Ecosystem Services? 79
5.4.7 Q7: Have You Ever Been Approached Before to Complete a Survey or an Interview? 79
5.5 Conclusions 79
References 80

6 Implications 81
6.1 Results for the Corridor 82
6.1.1 Error Analysis 84
6.2 Solutions 87
6.3 Lessons Learned 90
6.3.1 Sensitivity Analysis 92
6.4 Contributions 93
6.5 Conclusions 96
References 96

7 Conclusion . . . 97
7.1 Contribution . . . 97
7.2 Limitations . . . 98
7.3 Future Research . . . 100
7.4 Recommendations . . . 100
7.4.1 Applying the Footprint Calculator . . . 101
7.5 Concluding Remarks . . . 102
References . . . 103

Index . . . 105

About the Author

Mary J. Thornbush was a researcher of the Ecological Footprint Initiative based in the Faculty of Environmental Studies at York University, Canada until September 2020. She has over 80 publications in the areas of applied geomorphology and sustainability. Her doctoral thesis at the University of Oxford addressed urban sustainability through a study of air emissions from transport in central Oxford and their impacts on the weathering of its historical limestone buildings. Her relevant publications include a special journal issue on Geography, Urban Geomorphology and Sustainability in *Area* (2015) as well as books such as *Vehicular Air Pollution and Urban Sustainability: An Assessment from Central Oxford, UK* (2015, Springer) and a volume on *Urban Geomorphology: Landforms and Processes in Cities* (2018, Elsevier). She continues to collaborate on publications in the area of urban sustainability as well as most recently expanding her research interests to agricultural soils and rural sustainbility in southern Ontario.

Chapter 1
Introduction

Abstract This chapter conveys the conceptual and theoretical frameworks for the research. This includes intersecting literature on sustainability metrics (such as Ecological Footprint and biocapacity accounting), systems theory, and sustainability. The brief aims are centred around literature searches approached from a critical perspective to help inform the application and contribution of Ecological Footprint and biocapacity accounting as a sustainability metric. Therefore, in this chapter are criticisms directed at the National Footprint Accounts that are relevant to the Footprint Calculator. The focus here is on scale, aggregation, and data quality, including consideration of quality scores.

Keywords Sustainability · Natural/Social capital · Ecological Footprint and biocapacity accounting · National Footprint accounts · Footprint Calculator · Quality assurance · Quality scores · Criticisms · Issue themes · (Spatial) scale · Aggregation · Data quality

This chapter conveys the theoretical and conceptual frameworks for the research. This includes intersecting literature on sustainability metrics (such as Ecological Footprint and biocapacity accounting), systems theory, and sustainability. The brief aims centred around literature searches approached from a critical perspective to help inform the application and contribution of Ecological Footprint and biocapacity accounting as a sustainability metric.

Ecological Footprint and biocapacity accounting is intricate and requires time to understand the meaning of its concepts and overarching methodology, with its equations, constants, and idiosyncrasies. It has been heavily criticised in the academic literature, and this will be the basis of the critical approach adopted in this brief. This introductory chapter lays out the theoretical framework and rationale necessary to understand and justify the need for the Ecological Footprint and biocapacity as a sustainability metric. To begin, however, it is necessary to provide an overview of this system of biophysical accounts and give the reader the necessary background needed to both frame its application as well as its development since the

M. J. Thornbush, *The Ecological Footprint as a Sustainability Metric*,
SpringerBriefs in Environmental Science,
https://doi.org/10.1007/978-3-030-62666-2_1

early 1990s. In this way, the literature review presented in this chapter provides a context for its conceptualisation (in the conceptual framework) and methodology, including equations and data sources.

1.1 Overview

The Ecological Footprint and biocapacity accounting can be deployed to inform human consumption or demand set against the world's bioproductivity or supply of natural resources. It seeks to answer the question: 'How much of the regenerative capacity of the biosphere is being occupied by human activities?' (Wackernagel et al. 2007). Advocated in the early 1990s by William Rees and Mathis Wackernagel at the University of British Columbia, Canada (e.g., Rees 1992, 1996; Rees and Wackernagel 1996; Wackernagel and Rees 1996), the Ecological Footprint is considered to wholly represent the environmental dimension of sustainability.

In this brief, the author advocates that the Ecological Footprint does more than represent just the environmental dimension, even though it is a land-based composite indicator encompassing six land-type categories (crop land, grazing land, forest land, fishing grounds, built-up land, and carbon). As a consumption-based metric, the Ecological Footprint conveys information regarding consumer choices that impact trade and food security. As such, it has socioeconomic ramifications that need consideration additionally as part of the socioeconomic dimension of sustainability. Researchers have already, for instance, calibrated it against the Human Development Index (Lin et al. 2018), as a socioeconomic indicator of sustainable development, so that it can already reach these dimensions rather than just representing biophysical aspects.

Furthermore, this work contributes a case study application in the Alexander Skutch Biological Corridor of Costa Rica by considering the country's National Footprint Accounts in a time series since 1961. It is possible to track the overarching Ecological Footprint and biocapacity as well as individual land-based categories to augment understanding specific land types, such as crop land and carbon, influencing the country's overall Footprint. Finally, a critical examination of the methodology (both the National Footprint Accounts and the Footprint Calculator) appears last to clarify caveats in its application and the findings.

1.2 Theoretical Framework

Three main areas of concentration convey the theoretical frameworks that are most relevant to this study. Driving the Ecological Footprint as a sustainability metric is a 'full-world' model affecting the global carrying capacity as conveyed by Daly (1991, 2005, 2015). It is stipulated, for instance, that a world where resources are finite (according to a closed-loop or one-world model), is limited by its resource availability and constrained by planetary boundaries (Čuček et al. 2015; Fang et al.

2015; Steffen et al. 2015; Galli et al. 2016), or thresholds that restrain consumption. This provides the basis for the Ecological Footprint (e.g., as mentioned by Wackernagel and Rees 1996). For this reason, it is critical to understand the stocks and flows (Mancini et al. 2017) available for production that are used up in consumption and that generate wastes.

Systems theory conveys the interconnectedness of people to nature and the impacts of human-environment interactions. It recognises the relevance of scale in the operation of systems, including stocks (resources converted to goods or products) and flows (services), as recently acknowledged by the Global Footprint Network (e.g., Mancini et al. 2017). The research by Haberl et al. (2004) on the MEFA framework presents an exemplary (holistic) framework, that can be applied to address socioenvironmental (or 'sociobiophysical') systems, encompassing material and energy flows as well as economic growth and social well-being. It also denotes an integrated approach, such as found in social-ecological systems, that are relevant to this research. Systems theory also presents the possibility of knowledge-to-action and systemic intervention based on the interconnectedness of components within systems (cf. Midgley 2000).

The research focuses on a balanced approach within a sustainability framework that calls for environmental and socioeconomic components. This means that both the environment (as indicated by the land-based sustainability indicator represented by the Ecological Footprint and biocapacity accounting) and society – encompassing both economics as well as society at large – are both important factors to consider in integrated sustainability.

Therefore, the theoretical framework is based on the (three) components pertaining to the area of concentration entailed in this research. The components are:

1. **sustainability metrics**, as part of sustainability accounting, used to convey environmental performance through the balance of the Ecological Footprint and biocapacity;
2. **systems theory**, conveying the Earth as a closed system and, therefore, subject to a full-world model; and
3. **sustainability** as an integrated framework guiding social-ecological aspects of the research.

First, as a sustainability metric, the Ecological Footprint has been defined as an environmental composite indicator (cf. Strezov et al. 2017) that weighs human consumption (demand) against the average amount of land (global hectares or gha) necessary each year for production (supply) and to assimilate the generated waste (e.g., Wackernagel and Rees 1996). Highly developed (high-income) countries, for instance, are highly consumptive and, thereby, possess a large Ecological Footprint and demonstrate weak sustainability (cf. Wackernagel et al. 2006). This was encapsulated more recently by Lin et al. (2018), who conveyed a steadily increasing per person world Ecological Footprint at the expense of gradually declining world biocapacity (see their Figure 1, p 64).

Norström et al. (2014) noted three ways to enhance the likelihood of achieving the United Nations Sustainable Development Goals, among them adopting an integrated perspective, as represented by social-ecological systems (Virapongse et al. 2016; cf. Vasseur et al. 2017), as well as recognising cross-scalar interactions and thresholds. Accordingly, there is also the need to recognise the sociobiophysical context. These ambitions will form part of the proposed research by employing an integrated perspective that is outlined in the second and, particularly, third components of the research area of concentration (e.g., systems theory, sustainability) and by deploying the Ecological Footprint as a cross-scalar sustainability metric (e.g., Wackernagel et al. 1999).

The use of a sustainability assessment tool as the Ecological Footprint can benefit the determination of ecological deficit, for instance, which could impact ecosystems as well as energy and food security at the national level and, thereby, also influence human well-being (Sumaila et al. 2015). Ecological overshoot is occurring and affecting planetary health (cf. Čuček et al. 2015), indicating that ecological capital stocks are being depleted and that waste is accumulating (Borucke et al. 2013). This affects the planet's regenerative capacity as well as its ability to absorb wastes (Wackernagel et al. 2002). Authors, such as Kitzes et al. (2008), have proposed that this challenge be approached through resource management, such as of food and energy consumption, and spurred ecosystem production to augment the pool of natural capital.

Authors advocating a full-world model (e.g., Daly 2008) have already stressed the necessity of containing human economy (industry) and population. This includes stocks and flows of raw material inputs and waste sinks processed within the economic sphere (or 'ecosphere'). Such an ecological economic system operating in steady-state equilibrium would still encourage qualitative development, though not quantitative growth. In fact, authors are espousing 'strategic sustainable development' from a systems approach that endorses social-ecological sustainability within the boundaries or limits of the ecosphere (Robèrt et al. 2002).

Material flow accounting (e.g., material Footprint), for instance, considers the domestic extraction of natural resources as well as imports minus exports (Giljum et al. 2006). According to these authors, material flow accounting integrates financial resources and physical information within an accounting framework that can be used to inform policy. As a methodological approach, for instance, material flow accounting can also be deployed as an assessment tool for ecoefficiency policies. It draws on Daly's full-world model by setting the economy as a subsystem embedded within the environment (e.g. Costanza et al. 2014; also espoused by others from a socioeconomic perspective of the 'socioeconomic metabolism', e.g. Haberl et al. 2006, 2009, 2017). It relies on inputs of natural material and energy flow (asserted by Mancini et al. 2017 as natural capital stock and flows), depositing its wastes back into the natural system.

Such a metabolism approach recognises the organism as a system (Fischer-Kowalski and Hüttler 1998; Fischer-Kowalski and Haberl 2015) that is capable of undergoing change through sociometabolic transitions as part of sustainability transitions (e.g., Fischer-Kowalski 2011; Markard et al. 2012). This includes macro (landscape) transformations occurring across timescales of decades to centuries, as in the Viennese understanding of social-ecological transitions according to

Fischer-Kowalski and Rotmans (2009), comprising natural and societal metabolic processes. This was encapsulated by Haberl (2001, 2002; Haberl et al. 2007) as a system of ecological energy flows affecting net primary production (Fischer-Kowalski and Haberl 1998). The indicators involved can be grouped according to input, output, consumption, and trade, with trade indicators – including, for instance, domestic material consumption and total material consumption – relaying natural resource use and national consumption patterns (Giljum et al. 2006) and vice versa, since they are coevolving. These authors considered the Ecological Footprint as their second method of natural resource accounting as a sustainability measure.

In an ecosystems approach (see Figure 5 in Allen et al. 1993, p 13), ecosystems are impacted by humans from a full-world understanding that is necessarily multi-faceted. The spatial relationships between multiple species evident in communities are apparent in the landscape, while process and function focus on material and energy (matter-energy) flows. According to Allen et al. (1993), these fluxes are affected by decisions based on what is important at the time, so depend on specific spatial and temporal scales setting the context. The model is essentially biophysical because of its emphasis on interacting biota and environment in ecological systems set in a spatiotemporal context.

1.3 Conceptual Framework

The most critical concepts relevant to this research are the 'Ecological Footprint' and 'biocapacity'. The former is a measure of resource consumption comprising the demand for natural capital and the latter is natural capital itself measured through bioproductivity. Making up the Ecological Footprint are the six components of crop land, grazing land, forest land, fishing grounds, built-up land, and carbon as the carbon Footprint (cF – see Box 1.1 for relevant abbreviations after Lin et al. 2019). Carbon is the only waste-based component (Lin et al. 2019), while all remaining components of the Ecological Footprint are based on land productivity. Importantly, the Ecological Footprint has two main counterparts – that of consumption (EF_C) and production (EF_P), as denoted in Box 1.1. These terms also appear in the Glossary, that is based on Lin et al. (2019), and appears at the end of this brief. As exhaustively as possible, it conveys the important concepts necessary to comprehend this system of sustainability accounting.

Box 1.1 Acronyms

BC = Biocapacity
cF = carbon Footprint
CO_2 = Carbon Dioxide
CO_2-eq. = Carbon Dioxide Equivalents

(continued)

Box 1.1 (continued)
EF = Ecological Footprint
EFA = Ecological Footprint Analysis
EF_C = Ecological Footprint of Consumption
EF_P = Ecological Footprint of Production
EQF = Equivalence Factor(s)
FAO = Food and Agriculture Organization of the United Nations
FAOSTAT = Food and Agriculture Organization Corporate Statistical Database
GAEZ = Global Agro-Ecological Zones
GFN = Global Footprint Network
GHG = Greenhouse Gas(es)
gha = Global hectare
ha = Hectare
IEA = International Energy Agency (Paris, France)
IPCC = Intergovernmental Panel on Climate Change
IYF = Intertemporal Yield Factor(s)
NFA = National Footprint Accounts
NPP = Net Primary Production
t C = Tonne Carbon
UN COMTRADE = United Nations International Trade Statistics Database
wha = World average hectare

Source: Lin et al. (2019, p iv)

The Ecological Footprint and biocapacity accounting has been approached recently from a stock-flow perspective (Mancini et al. 2017, p 127). New definitions have been derived from this endeavour, as follows:

- **Biocapacity** – 'world-average hectare-equivalents (a stock-equivalent measure) that are available to provide the amount of ecosystem service flows actually regenerated by the biosphere every year'.
- **Ecological Footprint** – the 'hectare-equivalent units that are or would be needed to produce, in a year, the amount of ecosystem service flows humans actually use in that year'.

Given these new definitions, global hectares acquire a different meaning:

- **Global hectares (gha)** – are (still) not a measure of land use, but rather measure the inherent capacity of biosphere assets to produce useful biomass and services for appropriation by humans.

Understanding the difference between flows and stock in crop land, for instance, what is the natural capital in stock form that allows the growing of crops, and how to account for it, may address the problem of quality loss of the soil and the sustainability of management practices. A similar rationale may be applied to other land types, in consideration also of deforestation, the collapse of fisheries, and accumulation of CO_2 in the atmosphere.

1.3.1 Equations

The way that the Ecological Footprint is calculated is based on the GAEZ (Global Agro-Ecological Zones) model that ascribes a level of productivity to different biomes globally and so affects equivalence factors (EQF) used, themselves calculated through suitability indexes derived from GAEZ (Table 1.1 modified from Lin et al. 2019). For instance, the model considers crop land to be the most productive land-use type, followed by forest land, then shrub and grassland (or grazing land), and so on. There is an assumption that built-up land occupies formerly productive crop land. For this reason, its EF_C and BC (see Box 1.1) values are the same each year, as evident in the National Footprint Accounts 2020 Edition (Global Footprint Network 2020, using 2016 data) on which this study is based. It should be noted that equivalence values change with each edition of the National Footprint Accounts, thereby, setting the entire cross-temporal record (since 1961) for each new edition.

The main equations used are detailed in Box 1.2 based on Lin et al. (2019). Some of these equations depend on inputs either as data from a database (e.g., Eq. 6.2: World Resources Institute Global Land Cover Classification Database 2008) or estimates and specific figures derived from known sources (e.g. Eq. 6.2: Gulland 1971; Eq. 6.1: Pauly and Christensen 1995). Some of these sources are outdated and this will be considered later in Chap. 6 (Data Quality).

Table 1.1 EQFs for different land types

Land type	Equivalence factor (gha/wha)
Crop land	2.52
Grazing land	0.46
Forest land	1.29
Fishing grounds (Marine, Inland)	0.37
Built-up land (or Infrastructure)	2.52
Carbon	1.29

Modified from Figure 11.1 in Lin et al. (2019), p 54

Box 1.2 Equations

Equation 2.1: Calculation of Ecological Footprint of Consumption

$EF_C = EF_P + EF_1 - EF_E$

Corresponding with (the Footprint of consumption = production + imports – exports) associated with the product or waste

Equation 2.2: Calculation of biocapacity for Single Land Use Type

$BC = A * YF * IYF * EQF$

Biocapacity of a given land-use type = (its area within a country) (yield factor) (intertemporal yield factor) (equivalence factor)

Equation 2.3: Ecological Footprint of Yearly Product Extraction or Waste Generation

$EF_P = P/Y_N * YF * EQF * IY$

Ecological Footprint associated with a product or waste = (amount of product extracted or waste generated/national average yield for product extraction or waste generated) (yield factor of a given land-use type within a country) (equivalent factor for given land-use type) (intertemporal yield factor of a given land-use type)

Equation 2.3b: Ecological Footprint of Yearly Product Extraction or Waste Generation (Simplified)

$EF_P = P/Y_W * EQF * IY$

Substitute Y_W (world average yield for product extraction or waste absorption) for Y_N and remove *YF* from Eq. 2.3

Equation 2.4: Calculation for Yield of Derived Products

$Y^D = Y^P * EXTR$

Yield of derived product = (yield of parent product) (extraction rate)

Equation 6.1: PPR for Yield of Aquatic Species

$PPR = CC * DR * (1/TE)^{(TL - 1)}$

Primary production required = (carbon content of fish biomass = 1/9 according to Pauly and Christensen 1995) (discard rate of bycatch = 1.27, applied to all fish species, according to Pauly and Christensen 1995) (1/transfer efficiency of aquatic ecosystems = 10% according to Pauly and Christensen 1995) exp. (average fractional trophic level for given species – 1)

Equation 6.2: Yield for Aquatic Species

$Yield = 1/PPR * APP$

Yield for aquatic species or Y_A = (1/primary production requirement) (available primary productivity that can be sustainably harvested, based on the estimate by Gulland (1971), from each hectare of continental shelf area * *PPR* for world average harvested fish/area of continental shelf from World Resources Institute Global Land Cover Classification Database)

(continued)

Box 1.2 (continued)
Equation 10.1: Yield Factors Simple Calculation
$YF_N^L = Y_N^L / Y_W^L$
Yield factor for a given country and land type = yield for country and land type/world average yield for given land type

Equation 10.2: Yield Factors Extended Calculation
$YF_N^L = \sum A_W / \sum A_N$, where $A_N = P_N / Y_N$ and $A_W = P_N / Y_W$
Yield factor for a given country and land type = sum (area required to produce given product quantity using world average land)/sum (area harvested for a given product quantity in given country), where

Area harvested for a given product quantity in a given country = amount of given product extracted, or waste generated in a country/national yield for product extraction, and
Area required to produce a given product quantity using world average land = amount of given product extracted, or waste generated in country/world average yield for product extraction

Equation 10.3: Intertemporal Yield Factor Calculation
$IYF_{W,j} = \sum (P_{W,i,j} / Y_{W,j,b}) / \sum (P_{W,i,j} / Y_{W,i,j})$
World average intertemporal yield factor in year = sum (world amount of product harvested or CO_2 emitted for product *i* in year *j*/world product-specific yield for product *i* in the base year)/sum (world amount of product harvested or CO_2 emitted for product *i* in year *j*/world product-specific yield for product *i* in year *j*)

Source: Lin et al. (2019, Appendix A, pp. 63–66)

There are additional equations apparent in publications, as for example for the carbon component of the Ecological Footprint (refined more recently by Mancini et al. 2016):

$carbon\_EF = (P_C * S_{OCEAN}) / Y_W * EQF$, with $Y_W = AFCS/0.27$ and $AFCS = NFP/AF$, where:

- P_C is the world's annual anthropogenic emissions of CO_2 measured in Mt CO_2;
- S_{OCEAN} is the fraction of anthropogenic CO_2 emission that is sequestered by oceans in a given year. Data from Khatiwala et al. (2009) are currently used in National Footprint Accounts and the oceanic uptake fraction for the year 2010, for example, was 28% (Borucke et al. 2013);
- *EQF* is the equivalence factor used to weight forest land. (The GAEZ-based method assigns forest land a value, indicating that a hectare of world-average forest is that many times as productive as a world average hectare of land; see Table 1.1);
- Y_W is the annual rate of CO_2 sequestration per hectare of world average forest land;

- *AFCS* is the average forest carbon sequestration, expressed in t C ha^{-1} $year^{-1}$;
- *0.27* t C (t CO_2)$^{-1}$ represents the share of C within the CO_2 molecule, and is used to convert tonnes of carbon into tonnes of CO_2;
- *NFP* (Net Forest Production) is the total annual production of biomass in forests, expressed in tonnes of carbon per year (t C $year^{-1}$). Depending on the system boundaries under study, *NFP* can be defined as Gross Primary Production (GPP), Net Primary Production (NPP), Net Ecosystem Production (NEP) and Net Biome Production (NBP); and
- *AF* represents the total forested area on the Earth, expressed in hectares (ha).

1.3.2 Data Sources

The National Footprint Accounts are produced by the Global Footprint Network each year using the most recent data from the United Nations, as for example FAOSTAT and COMTRADE databases, reports available from the Intergovernmental Panel on Climate Change (IPCC), and data from the International Energy Agency (IEA). The most recent data for this study were available for 2016, and these have been used to determine the 2020 Edition of the National Footprint Accounts deployed in this brief. Importantly, details of the methodology, including software and protocols, are included in the working guidebook available from the Global Footprint Network (Lin et al. 2019). Data sources for the National Footprint Accounts are also conveyed in detail by Lin et al. (2019 – see their Appendix B, pp 67–71). There are several tables containing this information for the six components, livestock trade calculations, plus biocapacity. Because data sources have a significant impact on data quality, this will be considered next.

1.4 Quality Analysis

A critical approach adopted in this research forms the foundation of this section. Here, criticisms are addressed based on literature reviews that identify problems with the methods and data presented in this brief. These criticisms are rendered and summarised categorically as major themes. These themes include a focus on aggregation, spatial scale, and data quality. They draw from both issues stemming from the National Footprint Accounts as well as the Footprint Calculator. Besides presenting the criticisms, this section also conveys some recommendations for overcoming the shortcomings of the Ecological Footprint and biocapacity accounting based on a synthesis of the available literature, with added commentary by the author. One of the critical recommendations is for the incorporation of existing quality scores to help inform the margin of error in the National Footprint Accounts that would then affect the Footprint Calculator.

Before one can criticise the Footprint Calculator, it is necessary to inspect the National Footprint Accounts because they are weighted in the results at national level. In this section, both the Footprint Calculator and the National Footprint Accounts are necessarily approached from a critical perspective. First, the criticisms are presented based on four main themes (aggregation, spatial scale, energy-centrism, and data quality). The section ends with recommendations that in part justify the need for this study, but also reveal certain caveats and call for further research and enquiry.

1.4.1 Criticisms

The Footprint Calculator is based on the National Footprint Accounts at the national spatial scale. For this reason, it is necessary to consider issues with the National Footprint Accounts as well as the Footprint Calculator. These are already being considered by the Scientific Advisory Committee (SAC) of the Footprint Data Foundation, which is targeting the National Footprint Accounts. The author's original critical themes compiled from the literature appear in Table 1.2. These themes includes counts of the literature sources identified for each major theme based on the Zotero bibliographic database developed during the course of the author's role as Research Assistant to SAC during the term of this research at York University. By the end of her term, the database included 80/139 (almost 60%) annotated items according to Weaknesses as well as Strengths and suggestions. In these references, there were 240 instances where the themes in Table 1.2 were mentioned (see Table 1.2).

Personnel working for the National Footprint Accounts use Jira software as an issue tracker (or data cleaner – formerly Bugzilla). Items raised here are presented alongside a classification of the thematic issue(s) involved. A selection of (44) issues recorded in the Jira database was made on the basis that they were either representative of other issues addressed and/or raised substantial issues. Out of 305 issues currently recorded in Jira, this sample denotes a range of issues that are mostly either still 'To do' or 'In progress' (totalling 121 open issues), while some 184 issues are 'Done'.

Many of the issues recorded in the Jira database relate to calculations and updating sources, and these issues fall under Data quality (Table 1.3). Quality and Land use also appeared at least five times and represent issues associated with double-counting and potential causes for an inflated Ecological Footprint value. Issues such as Aggregation, 'False concreteness', Utility, and Energy-centrism are not specifically raised in the selection of 44 items from the Jira database. Nevertheless, these issues are recognised in the literature as problematic (see Table 1.2). However, they tend to be more overarching or general issues that cannot easily be easily remedied or resolved.

Table 1.2 Criticisms addressed by the literature arranged according to (10) emergent themes with some suggestions for improvement (in *italics*)

Major themes	Selection of sources
(1) **Aggregation** (45) – a simplistic and reductionist measure that uses one value for the Ecological Footprint; counteracted by six components that help to differentiate it; also, *a Footprint Family has been suggested*	Barrett et al. (2005), Borucke et al. (2013), Bossel (1999), Cranston et al. (2010), Curry and Maguire (2011), Erb (2004), Fang et al. (2015), Galli et al. (2011, 2012, 2016, 2020), Gasparatos et al. (2009), Giampietro and Saltelli (2014), Goldfinger et al. (2014), Hoekstra and Wiedmann (2014), Kharrazi et al. (2014), Kitzes and Wackernagel (2009), Kitzes et al. (2007, 2009), Lawn (2007), Lenzen and Murray (2001, 2003), Mayer (2008), McManus and Haughton (2006), Minx et al. (2009), Peters et al. (2008), Richardson (2019), Senbel et al. (2003), Singh et al. (2012), Solarin and Bello (2018), Troell et al. (2002), van den Bergh and Grazi (2014, 2015), van den Bergh and Verbruggen (1999), van Vuuren and Smeets (2000), Wackernagel (1998, 2014, 2019), Wackernagel and Yount (2000), Wackernagel et al. (2002), Wiedmann and Barrett (2010), Wiedmann and Minx (2008), Wiedmann et al. (2006, 2007) and Zhang et al. (2017)
(2) **Scale** (30): (2a) **Spatial scale** – arbitrary spatial scale used to calculate the Ecological Footprint, e.g. national boundaries affected by geopolitics and culture with no environmental meaning; *regions need to be defined from an environmental perspective, e.g. bioregions; some regions already used, e.g. for Asia*	Barrett et al. (2005), Blomqvist et al. (2013a, b), Borucke et al. (2013), Cranston et al. (2010), Erb (2004), Fang et al. (2015), Fiala (2008), Galli et al. (2012, 2016) and Gasparatos et al. (2009)
(2b) **Temporal scale** – a static measure that is incapable of making predictions, e.g. any potential effects on future loss of bioproductivity; there is a need to consider the long-term in sustainability, which *can be assessed for a specified edition of the National Footprint Accounts*	Giampietro and Saltelli (2014), Goldfinger et al. (2014), Haberl et al. (2001), Kitzes (2013), Kitzes and Wackernagel (2009), Kitzes et al. (2009), Lenzen and Murray (2001), Lenzen et al. (2007), Mayer (2008), McManus and Haughton (2006), Moffatt (2000), Senbel et al. (2003), Solarin and Bello (2018), Troell et al. (2002), van den Bergh and Verbruggen (1999), Wackernagel (2014), Wackernagel et al. (2004a) and Wiedmann et al. (2006, 2007)
(3) **'False concreteness'** (25) – creating false concreteness because of hypothetical rather than actual land use; *although real flows are used from real areas of land*	Blomqvist et al. (2013b), Erb (2004), Galli et al. (2011), Giampietro and Saltelli (2014), Goldfinger et al. (2014), Grazi et al. (2007), Haberl et al. (2001), Holmberg et al. (1999), Jóhannesson et al. (2020), Kharrazi et al. (2014), Kitzes and Wackernagel (2009), Kitzes et al. (2007), Lenzen and Murray (2003), Levett (1998), Mancini et al. (2017), Mayer

(continued)

Table 1.2 (continued)

Major themes	Selection of sources
	(2008), Monfreda et al. (2004), van den Bergh and Grazi (2014, 2015), van den Bergh and Verbruggen (1999), van Vuuren and Smeets (2000), Wackernagel (2014), Wackernagel and Yount (2000), Wackernagel et al. (2004a) and Wiedmann et al. (2006)
(4) **Utility** (3) – draws heavily from utility theory and an anthropocentric version of environmentalism; therefore, counts biocapacity only in terms of portions of the Earth which can be of direct use by people, e.g. biocapacity calculations exclude 36 billion hectares of land considered too unproductive – by excluding significant natural areas from estimates of biocapacity, the accounts do not recognise the interdependence of all ecosystems – from a systems approach	Kitzes et al. (2008), Lamb et al. (2014) and Venetoulis and Talberth (2008)
(5) **Quality** (33) – a measure of extensive production, but does not consider intensive production and its environmental impacts, e.g. land degradation	Blomqvist et al. (2013b), Borucke et al. (2013), Curry and Maguire (2011), Erb (2004), Fang et al. (2015), Fiala (2008), Galli et al. (2011, 2012, 2016), Giampietro and Saltelli (2014), Kharrazi et al. (2014), Kitzes and Wackernagel (2009), Kitzes (2013), Kitzes et al. (2007, 2008, 2009), Lenzen and Murray (2001, 2003), Lenzen et al. (2007), Mancini et al. (2017), McManus and Haughton (2006); Rees and Wackernagel (2013), Solarin and Bello (2018), Troell et al. (2002), van den Bergh and Grazi (2014), van den Bergh and Verbruggen (1999), Venetoulis and Talberth (2008), Wackernagel (1998, 2019), Wackernagel et al. (2004a, b) and Zhang et al. (2017)
(6) **Land use** (18) – single land-use functions are considered, when that may not be the reality, in order to avoid double-counting; however, neglect of multiple use can bias the Ecological Footprint upwards; furthermore, it is an incomplete environmental measure because it does not consider water use, persistent pollutants, and biodiversity	Borucke et al. (2013), Goldfinger et al. (2014), Holmberg et al. (1999), Kharrazi et al. (2014), Kitzes (2013), Kitzes and Wackernagel (2009), Lenzen et al. (2007), Mayer (2008), McManus and Haughton (2006), Peters et al. (2008), Troell et al. (2002), van den Bergh and Grazi (2015), van den Bergh and Verbruggen (1999), Wackernagel (2014), Wiedmann and Minx (2008), Wiedmann et al. (2006, 2007) and Wright et al. (2011)
(7) **Energy-centrism** (11) – the Ecological Footprint is dominated by energy, e.g. carbon Footprint – ecological overshoot is mostly attributable to the carbon Footprint, due to the hypothetical conversion of energy to land use, based on one strategy (reforestation) to assimilate wastes	Barrett et al. (2005), Fiala (2008), Galli et al. (2020), Giampietro and Saltelli (2014), Jóhannesson et al. (2020), McManus and Haughton (2006), Minx et al. (2009), Strezov et al. (2017), van den Bergh and Grazi (2014), van den Bergh and Verbruggen (1999) and Zhang et al. (2017)

(continued)

Table 1.2 (continued)

Major themes	Selection of sources
(8) **Equivalence factors** (21) – use of equivalence factors is problematic, e.g. if the EQF goes down 1 year, so does the Footprint because less biocapacity is assumed to be utilised; moreover, EQFs do not address large productivity differences within land-use types; various assumptions exist, e.g. that built-up land occupies productive land; *can estimate EQFs based on net primary production (NPP)*	Galli et al. (2011, 2012), Gasparatos et al. (2009), Goldfinger et al. (2014), Ferguson (2002), Haberl et al. (2001), Kitzes and Wackernagel (2009), Kitzes et al. (2007), Jóhannesson et al. (2020), Lenzen and Murray (2003), Lin et al. (2018), Monfreda et al. (2004), Singh et al. (2012), van Vuuren and Smeets (2000), Venetoulis and Talberth (2008), Wackernagel (1998), Wackernagel and Yount (2000) and Wackernagel et al. (2002, 2004a, b, 2019)
(9) **Yield factors** (18) – vast countries (such as Canada) can stretch over climatic zones; *could use local* yields	Ferguson (1999, 2002), Galli et al. (2011, 2012), Gasparatos et al. (2009), Goldfinger et al. (2014), Haberl et al. (2001), Jóhannesson et al. (2020), Kitzes et al. (2007), Lenzen and Murray (2001, 2003), Lenzen et al. (2007), Lin et al. (2018), Monfreda et al. (2004), Wackernagel (1998), Wackernagel and Yount (2000) and Wackernagel et al. (2004a, b)
(10) **Data quality** (36) – data (from official statistics) do not have an error margin, and cannot be quantified, e.g. land use, production, and consumption data are primarily from the FAO Statistical Database, International Energy Agency, and IPCC	Barrett et al. (2005), Blomqvist et al. (2013a), Clifton (2010), Collins and Flynn (2007), Curry and Maguire (2011), Ferguson (2002), Galli et al. (2016), Gasparatos et al. (2009), Giampietro and Saltelli (2014), Holmberg et al. (1999), Jóhannesson et al. (2020), Kitzes (2013), Kitzes and Wackernagel (2009), Kitzes et al. (2007), Lawn (2007), Lin et al. (2018), Mayer (2008), Mancini et al. (2016), McManus and Haughton (2006), Minx et al. (2009), Monfreda et al. (2004), Peters et al. (2008), Richardson (2019), Senbel et al. (2003), Troell et al. (2002), Venetoulis and Talberth (2008), Wackernagel (1998, 2014, 2019), Wackernagel and Yount (2000), Wackernagel et al. (2004a), Wiedmann and Minx (2008), Wiedmann et al. (2006, 2007), Wright et al. (2011) and Zhang et al. (2017)

Some of the main critical themes considered in this brief affecting the Footprint Calculator are summarised below:

- ***Aggregation** – in adding the consumption categories (food, housing, and transportation), it is not possible to see individual scores for these categories.
- ***(Spatial) Scale** – national information (from the National Footprint Accounts) is used to 'inform' the calculation of the Footprint Calculator.
- **Energy-centrism** – a country that uses 95% of renewable energy will affect household energy consumption qualitatively, although not necessarily quantitatively. There is also a cultural aspect to consider, as for example when heating and cooling are not extensively or intensively used.
- ***Data quality** – the survey needs to be culturally tailored for use, e.g. tropical countries.

Table 1.3 Frequencies of major issues appearing in the Jira database

Major theme (NFA-No.)	Frequency
Data quality	24
Land use	6
Quality	5
Yield factors	4
Equivalence factors	3
Scale (Spatial + Temporal)	2 (1 + 1)
Aggregation	0
'False concreteness'	0
Utility	0
Energy-centrism	0
Total	*44*

Three of these themes (*) will be detailed in the next sections of this chapter, including (1) (spatial) scale, (2) aggregation, and (3) data quality.

1.4.2 (Spatial) Scale

The literature indicates that spatial scale data, in particular, are problematic because of the 'arbitrary' boundaries that are imposed on the data. Some authors have suggested using 'bioregions' instead to address this issue. However, this continues to be an issue that cannot be easily resolved for the National Footprint Accounts. Spatial scale continues to also impact the Footprint Calculator (not only through the National Footprint Accounts, but also) through aggregation that overlooks specific locations, such as the brief's study area as a biological corridor. These areas cannot be construed equally to the national level, for example, because of the lifestyles lived by peasant farmers '*campesinos*' who reside in the Alexander Skutch Biological Corridor. By overlooking the particularities of this region, scale-specific cultural differences are overlooked, including regional environmental ethics in comparison to the national level. This flaw provides a basis for performing the fieldwork in the corridor and is a major rationale for this study.

1.4.3 Aggregation

The problem of aggregation has affected the data two-fold and, therefore, is another major issue that needs to be considered as being impactful on this study. First, aggregation occurs for the National Footprint Accounts when data are compiled at the national scale. The six categories of the National Footprint Accounts (crop land, grazing land, forest land, fishing grounds, built-up land, and carbon) can counteract this by dividing the data accordingly. However, they are still amalgamated at national level in the outputs for the Ecological Footprint and biocapacity and, hence, are susceptible to criticisms concerning national boundaries used rather than environmental units.

Second, for the Footprint Calculator, aggregation operates using the National Footprint Accounts to inform global Footprints, but also through spatial scaling from the individual or household level to town level (in the case study) and onwards from the local to the regional scale (for the corridor). This can be both positive and negative, and some have espoused that having a single (amalgamated) value in the Ecological Footprint can be beneficial (e.g., Giljum 2012; as essential for communication, Wiedmann and Barrett 2010). For example, having one value can simplify the results, which can then be deployed to inform governance (decision-making and policy-setting). On the other hand, there are many ways at which to arrive at a single value, and this can confound or simplify things, as the single value cannot convey the complexity involved.

1.4.4 Data Quality

Like the issue of Aggregation, Data quality impacts the National Footprint Accounts more than the Footprint Calculator. However, because the National Footprint Accounts are deployed in the Footprint Calculator, it is a relevant criticism. There are calculation-based issues that could affect Data quality (refer to Table 1.2). As shown in Table 1.3, the more frequent issue is associated with some aspect of 'Data quality'. This could be attributable to the quality of data reported to governments and input into United Nations datasets used in the National Footprint Accounts. Additionally, this issue could be compounded by whether information has been updated recently using recent sources and up-to-date information. The fact that data are reported with a lag time (e.g. 2020 Edition based on 2016 data) does not help to keep the editions of the National Footprint Accounts up-to-date.

As for the Footprint Calculator, Data quality could become an issue stemming from participants' understanding questions and properly answering them (without the aid of the interviewer, as would be the case of online participants – who are not guided in the process of answering the Footprint Calculator online survey). Furthermore, the quality of questions is an important consideration of any survey, including the questions posed by the Footprint Calculator. It can be challenged, for example, whether the questions (listed in Appendix 1.1) are exhaustive. Are questions pertaining to food, housing, and transportation the only questions necessary to fully determine the Ecological Footprint based on this survey? For example, the survey does not enquire about food quality (e.g., organic), although it does enquire about the consumption of processed foods. Regarding the use of renewable energy (Question 9), Costa Rica as an example uses 95% of renewables (and soon it will reach 100%), so this question will not be very variable for this country. Lastly, air transportation (Question 15) is not the norm in most developing countries, including Costa Rica, where participants may not fly every year – if ever.

Indeed, some of the questions in the Footprint Calculator are more effective at deciphering whether a person resides in a developed or developing country. For example, developing countries tend to be more dependent on what they produce (agriculture) and, therefore, are more likely to answer to consuming unprocessed, unpackaged, or locally grown food (Question 2). Similarly, the amount of garbage produced in a typical American household (in a developed country), for example, would be substantially greater than what a low- to medium-income/developing country would generate. As previously mentioned, the final question (Question 15) does not apply to most people in developing countries.

Although poverty is recognised to some degree in the questionnaire (e.g., housing materials in Question 4), many of the other questions are problematic for various reasons:

- Meat and seafood are expensive and may affect Question 1.
- Similarly, for Question 2, processed foods are less affordable.
- For Question 3, there may be running water in a house – in developed countries – but this does not mean that people can afford to pay for their water bills and, therefore, actually have running water.
- If they live in informal settlements, their house may be made from cheap/scrap materials (Question 4).
- Poor people may be unable to afford to buy contraception (unless publicly funded), so may have larger families (Question 5).
- Impoverished people will likely occupy smaller dwellings due to affordability (Question 6).
- Question 7 is problematic because although it asks whether there is electricity in the house (and possibly differentiates between developed or developing countries), it assumes that residents will use the electricity if they have it – which may not be the case, as for example in areas experiencing 'energy poverty'.
- Their homes may not be very energy efficient, again due to the affordability of modern appliances, etc. (Question 8).
- People who cannot afford to buy much food will not generate much trash (Question 10).
- Poverty may cause people to either carpool – perhaps with friends and family if there is no official carpool programme in place (Question 13), walk more, or cycle (if they can access a cycle) or, otherwise, use public transportation (Question 14).
- Very poor people cannot afford to fly (Question 15).

The survey assumes that people understand some concepts that uneducated people may never have come across before. For example, participants may not understand what it means to be 'energy efficient' in the home or otherwise. Likewise, they may not know the amount of renewable energy their house or country uses. Question 10 about garbage is, again, problematic because it compares the amount of garbage generated per household by comparison to neighbouring houses. But, what about isolated properties where people cannot see their neighbours, as in rural areas? For Question 12, participants may not know the fuel efficiency of vehicles – unless they are car-savvy. The survey can be construed as being more relevant to urbanites because of questions surrounding public transport, for instance, that may not be as relevant to country folk.

Importantly, there is an ignorance of 'tropical' climates in the survey. For example, Question 4 assumes that the type of housing material is based on preference, but may be based more on what is locally grown and readily available (e.g., straw/bamboo) – that is most appropriate for the climate. A country that has much clay, for instance in the UK, would use more bricks because clay is readily available and easy to acquire cheaply. The same can be said about using adobe in countries that deploy baked soil for building blocks. These 'preferences' could be attributed to climate and what can be locally sourced or, otherwise, to 'culture'. In fact, the Footprint Calculator survey is not very culturally informed – and culture constitutes more than a translation of the survey in different languages. In Costa Rica, for instance, people use fans as the only cooling implement and they do not heat their homes, as it is unnecessary in tropical countries. Therefore, the survey is climatically biased (e.g., 'temperate-centric'), and comes from the angle of an affluent urbanite (living in a wealthy nation) who can afford to buy things - as (net) consumers rather than producers.

1.4.4.1 Quality Scores

According to Lin et al. (2018, p 4), national results for each country are calculated based on as many as 15,000 data points from various datasets. A quality assurance (QA) process is used to validate these calculations. Quality scores are assigned as part of these QA stages. Finalised National Footprint Accounts are reviewed by the Global Footprint Network on a country-by-country basis in the last QA step, when quality scores are assigned that denote the reliability of these results (www.footprintnetwork.org/data-quality-scores). Accordingly, criteria for these quality scores are determined for each country in the National Footprint and biocapacity accounts. Each allotted quality score is made up of (1) a time series score, between 1 and 3 and (2) the latest year score, A to D (Table 1.4).

It is noteworthy from the website referring to quality scores (www.footprintnetwork.org/data-quality-scores) that, over time, it is possible to improve the quality of the results through collaborations with researchers, especially for those countries that have government agencies that can do so. For Costa Rica, as the national case study of focus in this brief, a score of 3A represents the highest quality possible outcome, where 'No component of BC or EF is unreliable or unlikely for any year'. This means that the National Footprint Accounts are most reliable here. This is important because the accounts are employed in the Footprint Calculator, as outlined in Chap. 4.

1.5 Brief Aims

This brief examines the Ecological Footprint and biocapacity accounting both from a critical approach of existing literature as well as its conceptualisation (the scientific basis) and methodology in application. The specific brief aims are:

Table 1.4 Criteria for qualifying National Footprint Accounts results

Score	Criteria
3A	No component of BC or EF is unreliable or unlikely for any year
3B	No component of BC or EF is unreliable or unlikely for the latest data year Some individual components of the EF or BC are unlikely in the latest data year The total EF and BC time series results are not significantly affected by unlikely data
3C	No component of BC or EF is unreliable or unlikely for the years prior to the latest data year Some individual components of the EF or BC are unlikely in the latest year Total EF and BC values are unlikely or unreliable in the most recent data year, but the ability to ascribe creditor/debtor status is unaffected in latest year
3D	No component of BC or EF is unreliable or unlikely for the years prior to the latest data year Some components of the EF or BC are very unlikely in the latest year EF and BC results in the latest year are significantly impacted by the unlikely or unreliable values, making them unusable
2A	EF or BC component time series have results that are very unreliable or very unlikely, except in the latest data year The total EF and BC time series results are not significantly affected by unlikely data No EF and BC results in the latest year are significantly affected by unlikely data
2B	EF or BC component time series have results that are very unreliable or very unlikely, including the latest year The total EF and BC time series results are not significantly affected by unlikely data
2C	Total EF or BC time series and component EF and BC time series results are unreliable or unlikely, especially in the latest year The total EF and BC time series results are not significantly affected by unlikely data The unlikely or unreliable values have most likely not impacted the creditor/debtor status in the latest year
2D	Total EF or BC time series and component EF and BC time series results are unreliable or unlikely, especially in the latest year The total EF and BC time series results are not significantly affected by unlikely data EF and BC results in the latest year are significantly impacted by the unlikely or unreliable values, making them unusable
1A	Several components of the EF or BC are very unreliable or unlikely, except the latest year The EF and BC time series results are significantly affected by unlikely data, and are unusable No EF and BC results in the latest year are significantly affected by unlikely data
1B	Several components of the EF or BC are very unreliable or unlikely, except the latest year The EF and BC time series results are significantly affected by unlikely data, and are unusable The total EF and BC results in the latest year are not significantly affected by unlikely data
1C	Several components of the EF or BC are very unreliable or unlikely The EF and BC time series results are significantly affected by unlikely data, and are unusable The unlikely or unreliable values have not impacted the creditor/debtor status
1D	There is too much unreliable or unlikely data to make any conclusions about the timeline or latest year of this country

Source: www.footprintnetwork.org/data-quality-scores

1. To present literature searches as background to the Ecological Footprint and biocapacity accounting from a critical approach to contextualise and justify the investigation.
2. To select a few key criticisms in a case study application to demonstrate the limitations and consider potential solutions based on Footprint Calculator outputs.
3. To consider the possible contributions that the Ecological Footprint and biocapacity accounting could make as a sustainability metric, despite its limitations.

These aims are foundational to this brief and provide a basis for its structure.

In the next chapters, the existing literature on the Ecological Footprint (Chap. 2) and biocapacity (Chap. 3) will be delineated to provide enough background information to readers, expanding on the theoretical and conceptual frameworks already presented in this chapter. A detailed case is presented for Costa Rica, including the methods (Chap. 4), based on farmers' survey-interviews completed in the study area. The findings are considered in comparison to the country's Ecological Footprint and biocapacity (Chap. 5). Error analysis pertaining to the case study application as well as solutions and lessons learned, including a sensitivity analysis based on the Footprint Calculator, form the bulk of Chap. 6. Finally, in Chap. 7, research limitations appear along with the contribution, recommendations, and suggestions for future research as part of the conclusion.

1.6 Conclusions

Amid the theoretical framework covered, this background information sets the stage for the remainder of the content. The conceptual framework presented in this first chapter is foundational for the remainder of this brief. In particular, the acronyms (included in this chapter) along with a glossary of the important terms are relayed as comprehensively as possible, including some important equations and data sources (that will be revisited later, in Chap. 6). Defined concepts included here offer a reference point for later chapters, including that of ecological overshoot. At the onset, however, it is important to understand that the Ecological Footprint and biocapacity accounting methodology encompasses two variables, that of consumption set against production at various scales.

Appendix 1.1: Footprint Calculator Survey (in English)

Food:

1. How often do you eat animal-based products? (beef, pork, chicken, fish, eggs, dairy products) Never/*Occasionally (really likes veggies – occasional meats, eggs/dairy)/Very often

   Add details to improve accuracy:

   - Beef or lamb? Pork? Poultry? Fish or shellfish? Eggs, cheese, and/or dairy?
     Often (nearly every day)

2. How much of the food that you eat is unprocessed, unpackaged or locally grown? (<320 kilometres/200 miles away) None/All/*30%

   Add details to improve accuracy:

   - How much of your diet is fresh, unpackaged foods?
   - How much of your diet is locally grown or produced? (<320 kilometres/200 miles away)

Housing:

3. Which housing type best describes your home? Freestanding, no running water/**Freestanding, running water**/Multi-storey apartment/*Duplex, row house or building with 2–4 housing units/Luxury condominium
4. What material is your house constructed with? Straw/bamboo/*Wood/**Brick or concrete**/Adobe/Steel or other
5. How many people live in your household? Just me/10+/*2
6. What is the size of your home? Tiny/*Medium (86 m^2 or 922 $feet^2$)/Huge
7. Do you have electricity in your home? *No/**Yes**
8. How energy efficient is your home? Hardly/*Average (modern appliances, climate controls)/Very
   [Information: How often do you run your heater or air conditioner? Do you use energy-efficient appliances and lighting? If you live in an energy-efficient house designed for passive heating and cooling or live in a mild climate where heating and cooling are unnecessary, adjust your answer towards 'very efficient'.]
9. What percentage of your home's electricity comes from renewable sources? (either directly or through purchased green power) Low/**High**/*12% (**95%**)
   [Information: About 14% of global primary energy consumption came from renewable, hydro, and nuclear in 2016, according to the BP Statistical Review of World Energy (June 2017).]

(continued)

10. Compared to your neighbours, how much trash do you generate? Much less/*Same/Much more

    Add details to improve accuracy:

    – What comes closest to your monthly new clothing, footwear, and/or sporting goods purchases? Shirts or underwear or socks/*Average
    – How often do you purchase new:

    Household appliances *Occasionally (I sometimes replace out-of-date appliances with new models)
    Electronics and gadgets *Occasionally (I replace out-of-date models and occasionally buy a new gadget)
    Books, magazines, and newspapers *Occasionally (I read some news online and subscribe to a couple of magazines or newspapers)
    What comes closest to your annual new household furnishings purchases? *Average (new bedding and a lamp or table, just to spruce things up)
    How much do you recycle: Plastic/Paper/*Some

Transportation:

11. How far do you travel by car or motorcycle each week? (as a driver or passenger)
    Car: Zero/Very far/*213 kilometres or 133 miles
    Moped: Zero/Very far/*5 kilometres or 3 miles
    [Information: If you only walk or cycle, choose zero. Both walking and bicycling reduce your Footprint and improve your well-being.]
12. What is the average fuel economy of the vehicles you use most often?
    Car: **Inefficient/*7 litres per 100 kilometres** or 34 miles per gallon/Efficient or electric
    Moped: **Inefficient *5 litres per 100 kilometres** or 45 miles per gallon/Efficient or electric
    [Information: Depending on the electricity source (% renewable versus % non-renewable), the fuel economy of a typical electric vehicle may range 2 to 6 litres per 100 kilometres or 40 to 150 miles per gallon.]
13. When you travel by car, how often do you carpool? **Never**/*60% – Occasionally/Always
14. How far do you travel on public transportation each week? (bus, train, etc.) Not very far/Very Far/*34 kilometres or 21 miles

    Add details to improve accuracy:

    – How far do you travel each week by train? *16 kilometres or 10 miles (**0 kilometres**)
    – How far do you travel each week by bus? *18 kilometres or 11 miles

(continued)

15. How many hours do you fly each year? **None**/Many/*6 hours

Transcribed by Mary Thornbush, Faculty of Environmental Studies, York University (Email: mthornbu@yorku.ca) based on: http://www.footprintcalculator.org/. (* = set default)

References

Allen TFH, Bandurski BL, King AW (1993) Ecosystem approach: theory and ecosystem integrity. International Joint Commission, Windsor. https://legacyfiles.ijc.org/publications/ID733.pdf

Barrett J, Birch R, Cherrett N, Wiedmann T (2005) Exploring the application of the Ecological Footprint to sustainable consumption policy. J Environ Policy Plan 7(4):303–316. https://doi.org/10.1080/15239080500441095

Blomqvist L, Brook BW, Ellis EC, Kareiva PM, Nordhaus T, Shellenberger M (2013a) Does the shoe fit? Real versus imagined Ecological Footprints. PLoS Biol 11(11):e1001700. https://doi.org/10.1371/journal.pbio.1001700

Blomqvist L, Brook BW, Ellis EC, Kareiva PM, Nordhaus T, Shellenberger M (2013b) The Ecological Footprint remains a misleading metric of global sustainability. PLoS Biol 11(11): e1001702. https://doi.org/10.1371/journal.pbio.1001702

Borucke M, Moore D, Cranston G, Gracey K, Iha K, Larson J, Lazarus E, Wackernagel M, Galli A (2013) Accounting for demand and supply of the biosphere's regenerative capacity: the National Footprint Accounts' underlying methodology and framework. Ecol Indic 24:518–533. https://doi.org/10.1016/j.ecolind.2012.08.005

Bossel H (1999) Indicators for sustainable development: theory, method, applications – a report to the Balaton group. IISD

Clifton D (2010) A sustainable world – an Ecological Footprint and I=PAT perspective. J Asia-Pac Centre Environ Account 16:3–26

Collins A, Flynn A (2007) Engaging with the Ecological Footprint as a decision-making tool: process and responses. Local Environ 12(3):295–312. https://doi.org/10.1080/13549830601183339

Costanza R, de Groot R, Sutton P, van der Ploeg S, Anderson SJ, Kubiszewski I, Farber S, Turner RK (2014) Changes in the global value of ecosystem services. Glob Environ Chang 26:152–158. https://doi.org/10.1016/j.gloenvcha.2014.04.002

Cranston GR, Hammond GP, Johnson RC (2010) Ecological debt: exploring the factors that affect national footprints. J Environ Policy Plan 12(2):121–140. https://doi.org/10.1080/15239081003719193

Čuček L, Klemeš JJ, Varbanov PS, Kravanja Z (2015) Significance of environmental footprints for evaluating sustainability and security of development. Clean Technol Environ 17 (8):2125–2141. https://doi.org/10.1007/s10098-015-0972-3

Curry R, Maguire C (2011) The use of Ecological and Carbon Footprint analysis in regional policy making: application and insights using the REAP model. Local Environ 16(9):917–936. https://doi.org/10.1080/13549839.2011.615306

Daly HE (1991) From empty world economics to full world economics: recognizing an historic turning point in economic development. In: Goodland R, Daly HE, El Serafy S, von Droste B (eds) Environmentally sustainable economic development: building on Brundtland. UNESCO, Paris. https://documents.worldbank.org/curated/pt/332821467989482335/pdf/multi-page.pdf

Daly HE (2005) Economics in a full world. Sci Am 293(3):100–107. https://www.jstor.org/stable/26061149

Daly HE (2008) Towards a steady-state economy. Essay commissioned by the Sustainable Development Commission, UK, April 24, 2008. https://is.muni.cz/el/1423/jaro2015/ENS242/um/55677449/3_Daly_2008_Towards_a_Steady_State_Economy.pdf

Daly HE (2015) Economics for a full world. Great Transition Initiative, June. https://www.greattransition.org/images/Daly-Economics-for-a-Full-World.pdf

Erb K (2004) Actual land demand of Austria 1926 – 2000. A variation on Ecological Footprint assessments. Land Use Policy 21(3):247–259. https://doi.org/10.1016/j.landusepol.2003.10.010

Fang K, Heijungs R, de Snoo GR (2015) Understanding the complementary linkages between environmental footprints and planetary boundaries in a footprint–boundary environmental sustainability assessment framework. Ecol Econ 114:218–226. https://doi.org/10.1016/j.ecolecon.2015.04.008

Ferguson ARB (1999) The logical foundations of ecological footprints. Environ Dev Sustain 1:149–156. https://doi.org/10.1023/A:1010070927485

Ferguson ARB (2002) The assumptions underlying eco-footprinting. Popul Environ 23 (3):303–313. https://doi.org/10.1023/A:1013051713184

Fiala N (2008) Measuring sustainability: why the ecological footprint is bad economics and bad environmental science. Ecol Econ 67(4):519–525. https://doi.org/10.1016/j.ecolecon.2008.07.023

Fischer-Kowalski M (2011) Analyzing sustainability transitions as a shift between socio-metabolic regimes. Environ Innov Soc Trans 1(1):152–159. https://doi.org/10.1016/j.eist.2011.04.004

Fischer-Kowalski M, Haberl H (1998) Sustainable development: socio-economic metabolism and colonization of nature. Int Soc Sci J 50(158):573–587. https://doi.org/10.1111/1468-2451.00169

Fischer-Kowalski M, Haberl H (2015) Social metabolism: a metric for biophysical growth and degrowth. In: Handbook of ecological economics. https://www.elgaronline.com/view/edcoll/9781783471409/9781783471409.00009.xml

Fischer-Kowalski M, Hüttler W (1998) Society's metabolism: the intellectual history of materials flow analysis, Part II, 1970–1998. J Ind Ecol 2(4):107–136. https://doi.org/10.1162/jiec.1998.2.4.107

Fischer-Kowalski M, Rotmans J (2009) Conceptualizing, observing, and influencing social–ecological transitions. Ecol Soc 14(2):3. https://doi.org/10.5751/ES-02857-140203

Galli A, Kitzes J, Wermer P, Wackernagel M, Niccolucci V, Tiezzi E (2011) An exploration of the mathematics behind the ecological footprint. WIT Trans State Art Sci Eng 51:249–256. https://doi.org/10.2495/ECO-V2-N4-250-257

Galli A, Wiedmann T, Ercin E, Knoblauch D, Ewing B, Giljum S (2012) Integrating Ecological, Carbon and Water footprint into a "Footprint Family" of indicators: definition and role in tracking human pressure on the planet. Ecol Indic 16:100–112. https://doi.org/10.1016/j.ecolind.2011.06.017

Galli A, Giampietro M, Goldfinger S, Lazarus E, Lin D, Saltelli A, Wackernagel M, Müller F (2016) Questioning the Ecological Footprint. Ecol Indic 69:224–232. https://doi.org/10.1016/j.ecolind.2016.04.014

Galli A, Iha K, Moreno Pires S, Mancini MS, Alves A, Zokai G, Lin D, Murthy A, Wackernagel M (2020) Assessing the Ecological Footprint and biocapacity of Portuguese cities: critical results for environmental awareness and local management. Cities 96:102442. https://doi.org/10.1016/j.cities.2019.102442

Gasparatos A, El-Haram M, Horner M (2009) The argument against a reductionist approach for measuring sustainable development performance and the need for methodological pluralism. Account Forum 33(3):245–256. https://doi.org/10.1016/j.accfor.2008.07.006

Giampietro M, Saltelli A (2014) Footworking in circles. Ecol Indic 46:260–263. https://doi.org/10.1016/j.ecolind.2014.06.019

Giljum S (2012) Integrating Ecological, Carbon and Water Footprint into a "Footprint Family" of indicators: definition and role in tracking human pressure on the planet. Ecol Indic 16:100–112. https://doi.org/10.1016/j.ecolind.2011.06.017

Giljum S, Hinterberger F, Wackernagel M, Kitzes J (2006) Resource use indicators in the European Union: policy processes, indicator sets and criteria for a headline indicator. Aachen Foundation, Aachen. www.aachenfoundation.org

Global Footprint Network (2020) National Footprint Accounts. https://data.footprintnetwork.org/

Goldfinger S, Wackernagel M, Galli A, Lazarus E, Lin D (2014) Footprint facts and fallacies: a response to Giampietro and Saltelli (2014) "Footprints to Nowhere". Ecol Indic 46:622–632. https://doi.org/10.1016/j.ecolind.2014.04.025

Grazi F, van den Bergh JCJM, Rietveld P (2007) Spatial welfare economics versus ecological footprint: modeling agglomeration, externalities and trade. Environ Resour Econ 38 (1):135–153. https://doi.org/10.1007/s10640-006-9067-2

Gulland JA (1971) The fish resources of the ocean. Fishing News, West Byfleet, Surrey. http://www.fao.org/3/a-al937e.pdf

Haberl H (2001) The energetic metabolism of societies Part I: accounting concepts. J Ind Ecol 5 (1):11–33. https://doi.org/10.1162/108819801753358481

Haberl H (2002) The energetic metabolism of societies Part II: empirical examples. J Ind Ecol 5 (2):71–88. https://doi.org/10.1162/10881980152830141

Haberl H, Erb K-H, Krausmann F (2001) How to calculate and interpret ecological footprints for long periods of time: the case of Austria 1926–1995. Ecol Econ 38(1):25–45. https://doi.org/10.1016/S0921-8009(01)00152-5

Haberl H, Fischer-Kowalski M, Krausmann F, Weisz H, Winiwarter V (2004) Progress towards sustainability? What the conceptual framework of material and energy flow accounting (MEFA) can offer? Land Use Policy 21:199–213. https://doi.org/10.1016/j.landusepol.2003.10.013

Haberl H, Krausmann F, Gingrich S (2006) Ecological embeddedness of the economy: a socioecological perspective on humanity's economic activities 1700-2000. Econ Polit Wkly 41(47):4896–4904. https://www.jstor.org/stable/4418950

Haberl H, Erb KH, Krausmann F, Gaube V, Bondeau A, Plutzar C, Gingrich S, Lucht W, Fischer-Kowalski M (2007) Quantifying and mapping the human appropriation of net primary production in earth's terrestrial ecosystems. PNAS 104(31):12942–12947. https://doi.org/10.1073/pnas.0704243104

Haberl H, Gaube V, Díaz-Delgado R, Krauze K, Neuner A, Peterseil J, Plutzar C, Singh SJ, Vadineanu A (2009) Towards an integrated model of socioeconomic biodiversity drivers, pressures and impacts. A feasibility study based on three European long-term socio-ecological research platforms. Ecol Econ 68(6):1797–1812. https://doi.org/10.1016/j.ecolecon.2008.11.013

Haberl H, Wiedenhofer D, Erb K-H, Görg C, Krausmann F (2017) The material stock–flow–service nexus: a new approach for tackling the decoupling conundrum. Sustainability 9(7):1049. https://doi.org/10.3390/su9071049

Hoekstra AY, Wiedmann TO (2014) Humanity's unsustainable environmental footprint. Science 344(6188):1114–1117. https://doi.org/10.1126/science.1248365

Holmberg J, Lundqvist U, Robèrt K-H, Wackernagel M (1999) The ecological footprint from a systems perspective of sustainability. Int J Sustain Dev World Ecol 6(1):17–33. https://doi.org/10.1080/13504509.1999.9728469

Jóhannesson SE, Heinonen J, Davíðsdóttir B (2020) Data accuracy in Ecological Footprint's carbon footprint. Ecol Indic 111(105983):9. https://doi.org/10.1016/j.ecolind.2019.105983

Kharrazi A, Kraines S, Hoang L, Yarime M (2014) Advancing quantification methods of sustainability: a critical examination emergy, exergy, ecological footprint, and ecological information-based approaches. Ecol Indic 37:81–89. https://doi.org/10.1016/j.ecolind.2013.10.003

Khatiwala S, Primeau F, Hall T (2009) Reconstruction of the history of anthropogenic CO_2 concentrations in the ocean. Nature 462:346–350. https://doi.org/10.1038/nature08526

Kitzes J (2013) An introduction to environmentally-extended input-output analysis. Resources 2 (4):489–503. https://doi.org/10.3390/resources2040489

Kitzes J, Wackernagel M (2009) Answers to common questions in Ecological Footprint accounting. Ecol Indic 9(4):812–817. https://doi.org/10.1016/j.ecolind.2008.09.014

Kitzes J, Peller A, Goldfinger S, Wackernagel M (2007) Current methods for calculating National Ecological Footprint Accounts. Sci Environ Sustain Soc 4:1):1–1):9

Kitzes J, Wackernagel M, Loh J, Peller A, Goldfinger S, Cheng D, Tea K (2008) Shrink and share: humanity's present and future Ecological Footprint. Philos T R Soc B 363(1491):467–475. https://doi.org/10.1098/rstb.2007.2164

Kitzes J, Moran D, Galli A, Wada Y, Wackernagel M (2009) Interpretation and application of the Ecological Footprint: a reply to Fiala (2008). Ecol Econ 68(4):929–930. https://doi.org/10.1016/j.ecolecon.2008.11.001

Lamb WF, Steinberger JK, Bows-Larkin A, Peters GP, Roberts JT, Wood FR (2014) Transitions in pathways of human development and carbon emissions. Environ Res Lett 9(1):014011. https://doi.org/10.1088/1748-9326/9/1/014011

Lawn P (2007) A stock-take of green national accounting initiatives. Soc Indic Res 80(2):427–460. https://doi.org/10.1007/s11205-006-0003-1

Lenzen M, Murray SA (2001) A modified ecological footprint method and its application to Australia. Ecol Econ 37(2):229–255. https://doi.org/10.1016/S0921-8009(00)00275-5

Lenzen M, Murray SA (2003) Ecological Footprint – issues and trends. The Ecological Footprint, 27 pp

Lenzen M, Hansson C, Bond S (2007) On the bioproductivity and land-disturbance metrics of the Ecological Footprint. Ecol Econ 61(1):6–10. https://doi.org/10.1016/j.ecolecon.2006.11.010

Levett R (1998) Footprinting: a great step forward, but tread carefully – a response to Mathis Wackernagel. Local Environ 3(1):67–74. https://doi.org/10.1080/13549839808725545

Lin D, Hanscom L, Murthy A, Galli A, Evans M, Neill E, Mancini MS, Martindill J, Medouar F-Z, Huang S, Wackernagel M (2018) Ecological Footprint accounting for countries: updates and results of the National Footprint Accounts, 2012–2018. Resources 7(3):58. https://doi.org/10.3390/resources7030058

Lin D, Hanscom L, Martindill J, Borucke M, Cohen L, Galli A, Lazarus E, Zokai G, Iha K, Eaton D, Wackernagel M (2019) Working guidebook to the National Footprint and biocapacity accounts. Global Footprint Network, Oakland. https://www.footprintnetwork.org/content/uploads/2019/05/National_Footprint_Accounts_Guidebook_2019.pdf

Mancini MS, Galli A, Niccolucci V, Lin D, Bastianoni S, Wackernagel M, Marchettini N (2016) Ecological Footprint: refining the carbon Footprint calculation. Ecol Indic 61:390–403. https://doi.org/10.1016/j.ecolind.2015.09.040

Mancini MS, Galli A, Niccolucci V, Lin D, Hanscom L, Wackernagel M, Bastianoni S, Marchettini N (2017) Stocks and flows of natural capital: implications for Ecological Footprint. Ecol Indic 77:123–128. https://doi.org/10.1016/j.ecolind.2017.01.033

Markard J, Raven R, Truffer B (2012) Sustainability transitions: an emerging field of research and its prospects. Res Policy 41:955–967. https://doi.org/10.1016/j.respol.2012.02.013

Mayer AL (2008) Strengths and weaknesses of common sustainability indices for multidimensional systems. Environ Int 34(2):277–291. https://doi.org/10.1016/j.envint.2007.09.004

McManus P, Haughton G (2006) Planning with Ecological Footprints: a sympathetic critique of theory and practice. Environ Urban 18(1):113–127. https://doi.org/10.1177/0956247806063963

Midgley G (2000) Systemic intervention: philosophy, methodology, and practice. Klumer Academic/Plenum Publishers, New York. https://www-springer-com.ezproxy.library.yorku.ca/gp/book/9780306464881

Minx JC, Wiedmann T, Wood R, Peters GP, Lenzen M, Owen A, Scott K, Barrett J, Hubacek K, Baiocchi G, Paul A, Dawkins E, Briggs J, Guan D, Suh S, Ackerman F (2009) Input-output analysis and carbon footprinting: an overview of applications. Econ Syst Res 21(3):187–216. https://doi.org/10.1080/09535310903541298

Moffatt I (2000) Ecological footprints and sustainable development. Ecol Econ 32(3):359–362. https://doi.org/10.1016/S0921-8009(99)00154-8

Monfreda C, Wackernagel M, Deumling D (2004) Establishing national natural capital accounts based on detailed Ecological Footprint and biological capacity assessments. Land Use Policy 21 (3):231–246. https://doi.org/10.1016/j.landusepol.2003.10.009

Norström A, Dannenberg A, McCarney G, Milkoreit M, Diekert F, Engström G, Fishman R, Gars J, Kyriakopoolou E, Manoussi V, Meng K, Metian M, Sanctuary M, Schlüter M, Schoon M, Schultz L, Sjöstedt M (2014) Three necessary conditions for establishing effective Sustainable Development Goals in the Anthropocene. Ecol Soc 19(3):8. https://doi.org/10.5751/ES-06602-190308

Pauly D, Christensen V (1995) Primary production required to sustain global fisheries. Nature 374:255–257. https://www.nature.com/articles/376279b0.pdf

Peters GM, Sack F, Lenzen M, Lundie S, Gellego B (2008) Towards a deeper and broader Ecological Footprint. P I Civil Eng-Eng Sustain 161(1):31–37. https://doi.org/10.1680/ensu.2008.161.1.31

Rees WE (1992) Ecological footprints and appropriated carrying capacity: what urban economics leaves out. Environ Urban 4(2):121–130. https://doi.org/10.1177/095624789200400212

Rees WE (1996) Revisiting carrying capacity: area-based indicators of sustainability. Popul Environ 17(3):195–215. https://doi.org/10.1007/BF02208489

Rees W, Wackernagel M (1996) Urban ecological footprints: why cities cannot be sustainable – and why they are a key to sustainability. Environ Impact Assess Rev 16(4):223–248. https://doi.org/10.1016/S0195-9255(96)00022-4

Rees WE, Wackernagel M (2013) The shoe fits, but the Footprint is larger than Earth. PLoS Biol 11(11):1–3. https://doi.org/10.1371/journal.pbio.1001701

Richardson RC (2019) Resource depletion is a serious problem, but "footprint" estimates don't tell us much about it. Greenbiz. https://www.greenbiz.com/article/resource-depletion-serious-problem-footprint-estimates-dont-tell-us-much-about-it

Robèrt K-H, Schmidt-Bleek B, Aloisi de Larderel J, Basile G, Jansen JL, Kuehr R, Thomas PP, Suzuki M, Hawken P, Wackernagel M (2002) Strategic sustainable development – selection, design and synergies of applied tools. J Clean Prod 10(3):197–214. https://doi.org/10.1016/S0959-6526(01)00061-0

Senbel M, McDaniels T, Dowlatabadi H (2003) The Ecological Footprint: a non-monetary metric of human consumption applied to North America. Glob Environ Chang 13(2):83–100. https://doi.org/10.1016/S0959-3780(03)00009-8

Singh RK, Murty HR, Gupta SK, Dikshit AK (2012) An overview of sustainability assessment methodologies. Ecol Indic 15(1):281–299. https://doi.org/10.1016/j.ecolind.2011.01.007

Solarin SA, Bello MO (2018) Persistence of policy shocks to an environmental degradation index: the case of Ecological Footprint in 128 developed and developing countries. Ecol Indic 89:35–44. https://doi.org/10.1016/j.ecolind.2018.01.064

Steffen W, Richardson K, Rockstrom J, Cornell SE, Fetzer I, Bennett EM, Biggs R, Carpenter SR, de Vries W, de Wit CA, Folke C, Gerten D, Heinke J, Mace GM, Persson LM, Ramanathan V, Reyers B, Sorlin S (2015) Planetary boundaries: guiding human development on a changing planet. Science 347(6223):1259855–1259855. https://doi.org/10.1126/science.1259855

Strezov V, Evans A, Evans TJ (2017) Assessment of the economic, social and environmental dimensions of the indicators for sustainable development. Sustain Dev 25:242–253. https://doi.org/10.1002/sd.1649

Sumaila UR, Hotte N, Galli A, Lam VWY, Cisneros-Montemayor AM, Wackernagel M (2015) Eco2: a simple index of economic–ecological deficits. Mar Ecol Prog Ser 530:271–279. https://doi.org/10.3354/meps11278

Troell M, Deutsch L, Rönnbäck P, Folke C, Kautsky N (2002) Misplaced critique: assumptions and applications of the EF, 10 pp

Van den Bergh JCJM, Grazi F (2014) Ecological Footprint policy? Land use as an environmental indicator. J Ind Ecol 18(1):10–19. https://doi.org/10.1111/jiec.12045

Van den Bergh JCJM, Grazi F (2015) Reply to the first systematic response by the Global Footprint Network to criticism: a real debate finally? Ecol Indic 58:458–463. https://doi.org/10.1016/j.ecolind.2015.05.007

Van den Bergh JCJM, Verbruggen H (1999) Spatial sustainability, trade and indicators: an evaluation of the 'ecological footprint'. Ecol Econ 29(1):61–72. https://doi.org/10.1016/S0921-8009(99)00032-4

Van Vuuren DP, Smeets EMW (2000) Ecological footprints of Benin, Bhutan, Costa Rica and the Netherlands. Ecol Econ 34(1):115–130. https://doi.org/10.1016/S0921-8009(00)00155-5

Vasseur L, Horning D, Thornbush M, Cohen-Shacham E, Andrade A, Barrow E, Edwards SR, Wit P, Jones M (2017) Complex problems and unchallenged solutions: bringing ecosystem governance to the forefront of the UN sustainable development goals. Ambio 46(7):731–742. https://doi.org/10.1007/s13280-017-0918-6

Venetoulis J, Talberth J (2008) Refining the ecological footprint. Environ Dev Sustain 10 (4):441–469. https://doi.org/10.1007/s10668-006-9074-z

Virapongse A, Brooks S, Metcalf EC, Zedalis M, Gosz J, Kliskey A, Alessa L (2016) A social-ecological systems approach for environmental management. J Environ Manag 178:83–91. https://doi.org/10.1016/j.jenvman.2016.02.028

Wackernagel M (1998) The ecological footprint of Santiago de Chile. Local Environ 3(1):7–25. https://doi.org/10.1080/13549839808725541

Wackernagel M (2014) Comment on "Ecological Footprint policy? Land use as an environmental indicator". J Ind Ecol 18(1):20–23. https://doi.org/10.1111/jiec.12094

Wackernagel M (2019) Ecological footprint accounting and its critics. GreenBiz. https://www.greenbiz.com/article/ecological-footprint-accounting-and-its-critics

Wackernagel M, Rees W (1996) Our Ecological Footprint: reducing human impact on the Earth. New Society Publishers, Gabriola Island

Wackernagel M, Yount JD (2000) Footprints for sustainability: the next steps. Environ Dev Sustain 2:21–42. https://doi.org/10.1023/A:1010050700699

Wackernagel M, Onisto L, Bello P, Linares AC, Falfán ISL, García JM, Guerrero AIS, Guerrero MGS (1999) National natural capital accounting with the ecological footprint concept. Ecol Econ 29(3):375–390. https://doi.org/10.1016/S0921-8009(98)90063-5

Wackernagel M, Schulz NB, Deumling D, Linares AC, Jenkins M, Kapos V, Monfreda C, Loh J, Myers N, Norgaard R, Randers J (2002) Tracking the ecological overshoot of the human economy. PNAS 99(14):9266–9271. https://doi.org/10.1073/pnas.142033699

Wackernagel M, Monfreda C, Erb K-H, Haberl H, Schulz NB (2004a) Ecological footprint time series of Austria, the Philippines, and South Korea for 1961–1999: comparing the conventional approach to an 'actual land area' approach. Land Use Policy 21(3):261–269. https://doi.org/10.1016/j.landusepol.2003.10.007

Wackernagel M, Monfreda C, Schulz NB, Erb K-H, Haberl H, Krausmann F (2004b) Calculating national and global Ecological Footprint time series: resolving conceptual challenges. Land Use Policy 21(3):271–278. https://doi.org/10.1016/j.landusepol.2003.10.006

Wackernagel M, Kitzes J, Moran D, Goldfinger S, Thomas M (2006) The Ecological Footprint of cities and regions: comparing resource availability with resource demand. Environ Urban 18 (1):103–112. https://doi.org/10.1177/0956247806063978

Wackernagel M, Wermer P, Goldfinger S (2007) Introduction to the Ecological Footprint: underlying research question and current calculation strategy. In: Internet encyclopaedia of ecological economics, pp 1–10. https://www.researchgate.net/publication/239582741_Introduction_to_the_Ecological_Footprint_Underlying_Research_Question_and_Current_Calculation_Strategy_Entry_prepared_for_the_Internet_Encyclopaedia_of_Ecological_Economics

Wackernagel M, Lin D, Evans M, Hanscom L, Raven P (2019) Defying the Footprint oracle: implications of country resource trends. Sustainability 11(7):2164. https://doi.org/10.3390/su11072164

Wiedmann T, Barrett J (2010) A review of the Ecological Footprint indicator – perceptions and methods. Sustainability 2(6):1645–1693. https://doi.org/10.3390/su2061645

Wiedmann T, Minx J (2008) A definition of ‘Carbon Footprint.’ In: Ecological economics research trends. Nova Science Publishers. https://www.novapublishers.com/catalog/product_info.php?products_id=5999

Wiedmann T, Minx J, Barrett J, Wackernagel M (2006) Allocating ecological footprints to final consumption categories with input-output analysis. Ecol Econ 56(1):28–48. https://doi.org/10.1016/j.ecolecon.2005.05.012

Wiedmann T, Lenzen M, Turner K, Minx J, Barrett J (2007) Multiregional input-output modelling opens new opportunities for the estimation of ecological footprints embedded in international trade. International Ecological Footprint conference, Cardiff, 8–10 May 2007, 22 pp

World Resources Institute Global Land Cover Classification Database (2008) Accessed 2008

Wright LA, Kemp S, Williams I (2011) ‘Carbon footprinting’: towards a universally accepted definition. Carbon Manage 2(1):61–72. https://doi.org/10.4155/cmt.10.39

Zhang L, Dzakpasu M, Chen R, Wang XC (2017) Validity and utility of ecological footprint accounting: a state-of-the-art review. Sustain Cities Soc 32:411–416. https://doi.org/10.1016/j.scs.2017.04.016

Chapter 2
The Ecological Footprint

Abstract Ecological Footprint and biocapacity accounting is approached in this brief from a systems perspective. Natural capital is seen as the foundation for all socioeconomic potential and action. Without nature, we are unable to capitalise on resources that are necessary to provide services and without which humanity would suffer. Using the Ecological Footprint, measured in in global hectares, it has been estimated that roughly 12 billion global hectares exist to support all life on Earth and human activities. If this is treated as Earth's carrying capacity, then it sets an upper boundary to resource exploitation. The two main threats to this life support system are human population size and consumption behaviour beyond basic life needs. Lavish lifestyles based on excessive consumption on a mass scale can cause problems of dwindling resources that need safeguarding. In this chapter, the literature that supports this framework is delineated to set the stage for the Ecological Footprint as a consumption-based composite indicator. The chapter conveys the evolution of this sustainability metric from a theoretical perspective that is grounded in the socioenvironmental as well as ecological economics in order to encapsulate an integrated approach from a systems perspective.

Keywords Natural capital · Ecological economics · Consumption · Systems theory · Ecological overshoot · Sustainability metrics · Composite indicator

Ecological Footprint and biocapacity accounting is approached in this brief from a systems perspective. Natural capital is seen as the foundation for all socioeconomic potential and action. Without nature, we are unable to capitalise on resources that are necessary to provide services, without which humanity would suffer. Using the Ecological Footprint measured in global hectares, it has been estimated that roughly 12 billion global hectares exist as highly productive land (Wackernagel 2014) to support all life on Earth and human activities. If this is treated as Earth's carrying capacity, then it sets an upper boundary for resource exploitation. The two main threats to this life support system are human population size and consumption behaviour beyond basic life needs. Lavish lifestyles based on excessive consumption

M. J. Thornbush, *The Ecological Footprint as a Sustainability Metric*,
SpringerBriefs in Environmental Science,
https://doi.org/10.1007/978-3-030-62666-2_2

on a mass scale can cause problems of dwindling resources that need safeguarding. In this chapter, the literature that supports this framework is delineated to set the stage for the Ecological Footprint as a consumption-based composite indicator. The chapter conveys the evolution of this sustainability metric from a theoretical perspective that is grounded in the socioenvironment as well as ecological economics in order to track the development of this composite indicator.

The Ecological Footprint is a composite indicator that comprises six components, including crop land, grazing land, forest land, fishing grounds, building-up land, and carbon. All components are consumption-based, except for the carbon Footprint that solely represents wastes. The consumption-based components are really production-based because production needs to precede consumption and the two are inevitably tied. For this, a systems approach is mandatory so that all components are seen connected to the whole – of the Ecological Footprint. In fact, one of the unique characteristics of the Ecological Footprint is its comparability to biocapacity in the determination of global ecological overshoot (Fang et al. 2015; Wackernagel 2014).

The Ecological Footprint itself cannot be separated from its components nor biocapacity, which is needed to support consumption. Biocapacity represents the natural production behind the resources needed for consumption to be possible. When there is a biocapacity deficit, consumption eventually suffers because it is circumscribed by limited resources. Although it is counterintuitive to consider the two (the Ecological Footprint and biocapacity) separately, for the sake of conveying different aspects of Ecological Footprint and biocapacity accounting, the two will be isolated in this and the next chapter (Chap. 3) in order to focus on each independently as different parts of the whole.

This chapter begins by relaying the evolution of the Ecological Footprint since its introduction in the early 1990s by Rees and Wackernagel. It proceeds by building the context for its development as a sustainability metric, specifically as a performance indicator. Lastly, it is placed within an integrated (systems) theoretical framework based on environmental and socioeconomic (sustainability) aspects most specific to ecological economics and emerging environmental accounts. This needs to be framed from a systems approach that is necessarily securely grounded in sustainability.

2.1 Ecological Footprint Evolution

The Ecological Footprint was established first conceptually before the methodology was more rigorously developed. Conceptually, the Ecological Footprint is part of an accounting system that contextualises it relative to biologically productivity (or 'bioproductivity') measured by biocapacity. As a land-based composite indicator (Borucke et al. 2013), the Ecological Footprint simply encompasses the amount of land required to sustain human consumption. The accounting system, namely Ecological Footprint and biocapacity accounting, is based on global hectares (gha) that represent a hectare calibrated using an equivalence factor to ensure comparability at global scale. In addition, a yield factor is also deployed in its calculation to account for different levels of productivity around the world affected by natural (e.g., climate) and human-induced factors (e.g., soil fertility). According to Jóhannesson et al. (2020, p 3), for example, two coefficients are deployed to convert areal size from hectares to global hectares: equivalence factors (EQF) account for differences in productivity between land types; and yield factors (YF) account for different productivity between the same land types but in different countries. In this way, local–regional Footprints and biocapacity can be compared to world-average productivity.

By comparing the Ecological Footprint to biocapacity results, it is possible to decipher whether nations are 'in the red' or 'in the green', respectively meaning that they are either in ecological deficit (EF > BC) or have an ecological reserve (EF < BC). This is tied to 'ecological overshoot' through the persistent excessive use of biocapacity relative to the resources available to support human life and activities. This can be assessed at various scales because the Ecological Footprint is known to be a cross-scalar measure (cf. Wackernagel 2014). This reflects human impacts at different scales, from the local – including the built environment – to regional and global levels. It is possible to attain such results based on the level of interest for aggregating results for the Ecological Footprint and biocapacity. Most often, however, the national scale is employed, unless the Footprint Calculator is used to assess individual–household or community level (to the regional) scales, as in the current study. The Footprint Calculator, which will be addressed in more detail later (Chap. 4), is a global Footprint calculator with inputs from the National Footprint Accounts.

The National Footprint Accounts, therefore, are an important resource available from the Global Footprint Network to access information at national–regional level. There are two main tools available from the Global Footprint Network, the National Footprint Accounts and Footprint Calculator, with the former appearing as a data explorer for over 200 countries (see Open Data Platform: https://data.footprintnetwork.org/#/). The edition of the accounts used in this brief is the 2020 Edition, which is based on data from 2016. There is much published based on the National Footprint Accounts, and some of these publications appear – usually as methodological updates – each year with new editions (e.g. most recently, Lin et al. 2018). In addition, a Working guidebook to the National Footprint and biocapacity accounts (Lin et al. 2019) is updated each year as a 'manual' to the accounts. By comparison, the Footprint Calculator is a popular instrument for assessing Footprints, but it has not been scrutinised to the same extent that the National Footprint Accounts have been examined (with the exception of work on the contributions of personal Footprint calculators by Collins et al. 2018, 2020). Perhaps this is the next logical step in Footprint evolution.

A recent focus has been on using Earth Overshoot Day to build awareness and work towards sustainability. This uses the yearly calendar to pinpoint where different countries are positioned relative to resource consumption based on a date that indicates when their resources for the entire year have been consumed. The earlier in the year for Earth Overshoot Day, the worse off a nation is in terms of its sustainability prospects. High-income countries, such as Canada and the United States, have their Overshoot Days in the first half of the year. For example, in 2019, Canada's Overshoot Day was March 18, 2019 relative to the most recent global Earth Overshoot Day of July 29, 2019 (https://www.overshootday.org/). The Earth Overshoot Day was delayed by 3 weeks because of the coronavirus (COVID-19) in 2020 (https://www.overshootday.org/newsroom/country-overshoot-days/). By comparison, less developed (low- to medium-income) countries appear later in the year, as for instance Costa Rica had an Overshoot Day on August 10, 2020 (https://www.overshootday.org/). Whether a country is in a position of 'weak' or 'strong' sustainability could be quantified using the month that Overshoot Day occurs; this can also be conveyed at the global scale. Such an approach also fosters a sense of sustainability in the measure because of the way that it is used. More specifically, by using time as an indicator of resource availability versus depletion, it is possible to stimulate longer term thinking and planning. Perhaps such an increased awareness since 2010 has been impacting the human expenditure of resources and augmented conservation efforts.

An essential way that the Ecological Footprint has emerged has been as part of sustainability. 'Weak' sustainability is measured when deficits are evident (Galli et al. 2018; Pearce and Atkinson 1993; Wackernagel et al. 2006), for instance, whereas 'strong' sustainability can be achieved when reserves exist and are maintained to put off going into a situation of ecological overshoot, especially at the global scale. Nationally, at least, it is still possible to exchange resources through traded commodities. However, when humanity reaches a situation of global overshoot, competition for resources is augmented and even resource exchange could not overcome issues of resource depletion in circumstances of weak sustainability. Trade has, therefore, become increasingly subject in the work produced by the Global Footprint Network. Part of this address has been because of criticisms of an 'antitrade sentiment' inherent in the methodology (e.g., McManus and Haughton 2006; van den Bergh and Grazi 2014; van den Bergh and Verbruggen 1999 – Chap. 6 addresses such criticisms in more detail).

To tackle this, a systems approach is increasingly recognised, where inputs from foreign trade into countries counterweigh outgoing resources traded out of countries. It is important to realise that importing from other nations is not an ideal situation for nations. The reason for this is that it weakens economies that are dependent on other countries and subject to prices in the global market. This consideration has led to comparisons between countries regarding their current and projected positions globally. Nations that are self-sufficient and not reliant on others (or dependent) for traded inputs are in stronger positions to sustain their economies because their resources can support their populations (cf. Wackernagel and Beyers 2019). Such nations, including Canada, have abundant resources – either because of their greater natural inheritance of biocapacity or because their populations are not depleting their resource base (Wackernagel and Beyers 2019) at a fast rate, so that the land can continue to regenerate and support the population. By contrast, island nations that have a small landmass – as for example Japan, the United Kingdom, and so on – are more limited by their natural capital and reliance on trade.

2.2 Sustainability Metrics

In order to assess our current status and progress in attaining sustainability at any scale, it is necessary to be able to qualify it using performance indicators. The Ecological Footprint is one of such indicators that enables us to ascertain whether we are living sustainably within the Earth's limits. Carrying capacity will be addressed in the next chapter (Chap. 3) and is one of the essential aspects that needs to be determined at various scales, including at the planetary level. This has been an area of research interest recently, as planetary health and the boundaries needed to attain this are being addressed (Čuček et al. 2015; Fang et al. 2015; Häyhä et al. 2016; O'Neill et al. 2018).

The Ecological Footprint is a composite indicator that includes different components (crop land, grazing land, forest land, fishing grounds, built-up land, and carbon), including five that are based on production and only one (carbon) that is a waste component (Fig. 2.1, based on the Global Footprint Network 2020). Built-up land assumes that urbanisation is taking up previously productive land that could have been used for agriculture (as crop land). Criticisms have targeted that the Ecological Footprint only considers anthropogenic land use (utility theory, e.g. Venetoulis and Talberth 2008) and does not recognise non-productive land, although this is considered in the built-up land component – although under the assumption that cities are occupying formerly productive land.

As an environmental indicator, the Ecological Footprint only indirectly considers human activities and behaviour. Unlike the Human Development Index, which is based on three variables, namely life expectancy, per person gross domestic product (GDP), and a combined educational indicator, consisting of adult literacy rates and primary, secondary, and tertiary school enrolment rates. However, this

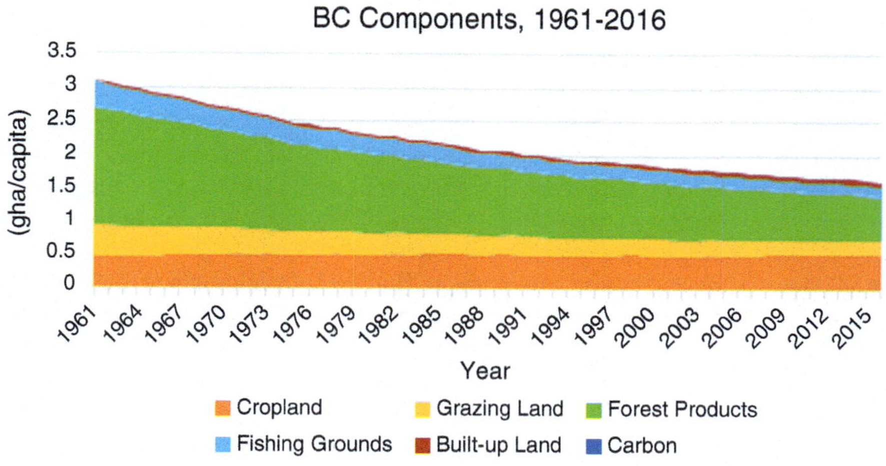

Fig. 2.1 Components of the world's Ecological Footprint, 1961–2016. (Source: Global Footprint Network 2020)

socioeconomic metric lacks the ability to integrate the environmental dimension that the Ecological Footprint focuses on as a measure (Strezov et al. 2017). The Ecological Footprint has been correlated to the Human Development Index in order to check its performance against a human-based index (e.g., one that considers socioeconomics). The research by Lin et al. (2018) concluded that countries that score high on the Human Development Index have a high Footprint, so that there is a direct (positive) correlation between the two measures. This is mainly attributable to the carbon Footprint that has been observed (as the 'energy Footprint') to convey wealth, even at household level (cf. Minx et al. 2009). Although this linkage can broaden the human dimension being considered by the Ecological Footprint, it does not on its own merit represent a full integration of this aspect. Arguably, this can be executed through the questions in the Footprint Calculator that will be presented in Chap. 4.

2.3 Systems Approach

The Ecological Footprint and biocapacity accounting is a unique methodology because of its comparison of human consumption with Earth's natural capital, its ability to assimilate wastes, and regeneration capacity. It inherently adopts a systems approach that considers inputs and outputs as well as flows and stocks or components in the system (e.g., Borucke et al. 2013; Mancini et al. 2017; Niccolucci et al. 2009, 2011). It has been conceptualised as a simple input-output (I-O) system, where inputs include the natural stocks that support human life and activities. Human outputs are also considered in the system through its components that include wastes from production, such as CO_2 emissions from manufacturing. It is possible to derive stocks from the available flows of global hectares per year.

As a cross-scalar metric, the methodology allows for calculations of the Ecological Footprint and biocapacity at various scales, from the individual (household) level using the Footprint Calculator, to the regional scale with accumulation of responses in a bottom-up approach – for instance, as part of a multiregional input-output (MRIO) model. Otherwise, it is possible to use the National Footprint Accounts in a top-down approach that allows for national–global level calculations. The former (bottom-up) approach provides contextualisation based on local influences, such as cultural aspects that influence the balance between nature's production (biocapacity) and human consumption. However, the information is acquired in a top-down approach based on national government data. The ecosystems approach (cf. Allen et al. 1993) is appropriate to encapsulate the multi-scalar characteristic of the methodology because of its similar multilevel approach or multifaceted approach to landscapes.

2.4 Integrated Sustainability

The integrated human-nature interactions captured by the methodology allows for a social-ecological systems approach that combines ecological and economic sustainability (Common and Perrings 1992). Trade is an example of how the latter (economics) is considered in the Ecological Footprint and biocapacity accounting. Systems thinking is relevant here, as it concerns a holistic approach to both simple and more complex relationships between different subsystems (Ostrom 2009), including the work by Daly (e.g., 2008) who considered natural resources to be a subsystem supporting the economic sector (another subsystem) in the ecosphere. It is integrative, furthermore, as it is relevant to all three components of environment and socioeconomics. For example, Virapongse et al. (2016) relayed the potential of social-ecological systems integrated between the natural and social sciences, as for example to address management issues through applied transdisciplinarity and the concept of social-ecological resilience (cf. Adger et al. 2005; Max-Neef 2005). Another systems framework by Fischer et al. (2015) shows social-ecological systems across scales (landscape, regional, and global) in addition to temporal dynamics, as part of a multifaceted ecosystem approach (cf. Allen et al. 1993).

There are various applications of sustainability, as for instance as environmental sustainability that focuses on the use of natural resources (Lozano 2008), allowing for interdisciplinary engagement with the development of the concept as it emerges. An integrated approach is preferred (e.g., for resilient dynamic, adaptive complex systems, cf. Fiksel 2006; Folke 2006). Although defining it can be challenging, according to Costanza and Patten (1995), the general notion of sustainability recognises that 'a sustainable system is one which survives and persists'. The context of course is important, as are the spatial-temporal considerations that should accompany any definition of sustainability (Brown et al. 1987).

Like sustainability, systems thinking embraces an interdisciplinary perspective from an integrated sustainability framework for the environmental and socioeconomic dimensions of human consumption impacting the Ecological Footprint. Midgley (2000), for instance, developed a systemic intervention model that incorporates judgement, critique, and action. The latter component is especially interesting because of its emphasis on 'action for improvement' (p 132). This can benefit from a holistic approach, such as that of social-ecological systems operating within overlapped natural and cultural spheres of causation, encompassing biophysical actualities (of the material world) that are socially relevant (e.g., Vasseur et al. 2017). The association between parts emphasises relationships in the systems perspective and is, thereby, considered to be part of systems science as well as an 'organismic approach' (Hammond 2003), with the living organism denoting an open system (von Bertalanffy 1950). This conveys the notion of a 'layered structure' (Checkland 1999), but also different scales of operation within a system – as with subsystems (Ostrom 2009), conveying the organized nature of the whole based on relationships between its parts.

2.5 Conclusions

Although it has been strictly classified as an environmental composite indicator (Strezov et al. 2017), the Ecological Footprint and biocapacity accounting methodology integrates human activities, as through its consideration of economic factors like trade. To be an integrated sustainability metric, it needs to consider the environmental and socioeconomic dimensions as well as entail the cultural domain that may impact other aspects. The Ecological Footprint, as the demand-side of the methodology, encapsulates human lifestyles influencing resource use and the incumbent generation of wastes. In this way, it functions as a simple I-O system, where flows are controlled by the pull from human demand to consume resources. This is affected by the level of societal development, but also by population size and other factors. The next chapter (Chap. 3) of this brief examines carrying capacity with a focus on biocapacity.

References

Adger WN, Hughes TP, Folke C, Carpenter SR, Rockström J (2005) Social-ecological resilience to coastal disasters. Science 309:1036–1039. https://doi.org/10.1126/science.1112122

Allen TFH, Bandurski BL, King AW (1993) Ecosystem approach: theory and ecosystem integrity. International Joint Commission, Windsor. https://legacyfiles.ijc.org/publications/ID733.pdf

Borucke M, Moore D, Cranston G, Gracey K, Iha K, Larson J, Lazarus E, Wackernagel M, Galli A (2013) Accounting for demand and supply of the biosphere's regenerative capacity: the National Footprint Accounts' underlying methodology and framework. Ecol Indic 24:518–533. https://doi.org/10.1016/j.ecolind.2012.08.005

Brown BJ, Hanson ME, Liverman DM, Merideth RW Jr (1987) Global sustainability: toward definition. Environ Manag 11(6):713–719. https://doi.org/10.1007/BF01867238

Checkland P (1999) Systems thinking. Chapter 3. In: Currie WL, Galliers B (eds) Rethinking management information systems. Oxford University Press, New York, pp 45–56

Collins A, Galli A, Patrizi N, Pulselli FM (2018) Learning and teaching sustainability: the contribution of ecological footprint calculators. J Clean Prod 174:1000–1010. https://doi.org/10.1016/j.jclepro.2017.11.024

Collins A, Galli A, Hipwood T, Murthy A (2020) Living within a one planet reality: the contribution of personal footprint calculators. Environ Res Lett 15(2):025008. https://doi.org/10.1088/1748-9326/ab5f96

Common M, Perrings C (1992) Towards an ecological economics of sustainability. Ecol Econ 6:7–34. https://doi.org/10.1016/0921-8009(92)90036-R

Costanza R, Patten BC (1995) Defining and predicting sustainability. Ecol Econ 15(3):193–196. https://doi.org/10.1016/0921-8009(95)00048-8

Čuček L, Klemeš JJ, Varbanov PS, Kravanja Z (2015) Significance of environmental footprints for evaluating sustainability and security of development. Clean Technol Envir 17(8):2125–2141. https://doi.org/10.1007/s10098-015-0972-3

Daly HE (2008) Towards a steady-state economy. Essay commissioned by the Sustainable Development Commission, UK (April 24, 2008). https://is.muni.cz/el/1423/jaro2015/ENS242/um/55677449/3_Daly_2008_Towards_a_Steady_State_Economy.pdf

Fang K, Heijungs R, de Snoo GR (2015) Understanding the complementary linkages between environmental footprints and planetary boundaries in a footprint–boundary environmental

sustainability assessment framework. Ecol Econ 114:218–226. https://doi.org/10.1016/j.ecolecon.2015.04.008
Fiksel J (2006) Sustainability and resilience: toward a systems approach. Sustain Sci Pract Pol 2 (2):14–21. https://doi.org/10.1080/15487733.2006.11907980
Fischer J, Gardner TA, Bennett EM, Balvanera P, Biggs R, Carpenter S, Daw T, Folke C, Hill R, Hughes TP, Luthe T, Maass M, Meacham M, Norström AV, Peterson G, Queiroz C, Seppelt R, Spierenburg M, Tenhunen J (2015) Advancing sustainability through mainstreaming a social–ecological systems perspective. Curr Opin Env Sust 14:144–149. https://doi.org/10.1016/j.cosust.2015.06.002
Folke C (2006) Resilience: the emergence of a perspective for social-ecological systems analyses. Global Environ Chang 16(3):253–267. https://doi.org/10.1016/j.gloenvcha.2006.04.002
Galli A, Đurović G, Hanscom L, Knežević J (2018) Think globally, act locally: implementing the sustainable development goals in Montenegro. Environ Sci Pol 84:159–169. https://doi.org/10.1016/j.envsci.2018.03.012
Global Footprint Network (2020) National Footprint Accounts. https://data.footprintnetwork.org/#/
Hammond D (2003) The science of synthesis: exploring the social implications of general systems theory. University Press of Colorado, Boulder. http://www.jstor.org.ezproxy.library.yorku.ca/stable/j.ctt1d8h9jg
Häyhä T, Lucas PL, van Vuuren DP, Cornell SE, Hoff H (2016) From planetary boundaries to national fair shares of the global safe operating space — how can the scales be bridged? Global Environ Chang 40:60–72. https://doi.org/10.1016/j.gloenvcha.2016.06.008
Jóhannesson SE, Heinonen J, Davíðsdóttir B (2020) Data accuracy in Ecological Footprint's carbon footprint. Ecol Indic 111:105983:9. https://doi.org/10.1016/j.ecolind.2019.105983
Lin D, Hanscom L, Murthy A, Galli A, Evans M, Neill E, Mancini MS, Martindill J, Medouar F-Z, Huang S, Wackernagel M (2018) Ecological footprint accounting for countries: updates and results of the National Footprint Accounts, 2012–2018. Resources 7(3):58. https://doi.org/10.3390/resources7030058
Lin D, Hanscom L, Martindill J, Borucke M, Cohen L, Galli A, Lazarus E, Zokai G, Iha K, Eaton D, Wackernagel M (2019) Working guidebook to the National Footprint and biocapacity accounts. Global Footprint Network, Oakland. https://www.footprintnetwork.org/content/uploads/2019/05/National_Footprint_Accounts_Guidebook_2019.pdf
Lozano R (2008) Envisioning sustainability three-dimensionally. J Clean Prod 16:1838–1846. https://doi.org/10.1016/j.jclepro.2008.02.008
Mancini MS, Galli A, Niccolucci V, Lin D, Hanscom L, Wackernagel M, Bastianoni S, Marchettini N (2017) Stocks and flows of natural capital: implications for ecological footprint. Ecol Indic 77:123–128. https://doi.org/10.1016/j.ecolind.2017.01.033
Max-Neef MA (2005) Foundations of transdisciplinarity. Ecol Econ 53(1):5–16. https://doi.org/10.1016/j.ecolecon.2005.01.014
McManus P, Haughton G (2006) Planning with ecological footprints: a sympathetic critique of theory and practice. Environ Urban 18(1):113–127. https://doi.org/10.1177/0956247806063963
Midgley G (2000) Systemic intervention: philosophy, methodology, and practice. Klumer Academic/Plenum Publishers, New York. https://www-springer-com.ezproxy.library.yorku.ca/gp/book/9780306464881
Minx JC, Wiedmann T, Wood R, Peters GP, Lenzen M, Owen A, Scott K, Barrett J, Hubacek K, Baiocchi G, Paul A, Dawkins E, Briggs J, Guan D, Suh S, Ackerman F (2009) Input-output analysis and carbon footprinting: an overview of applications. Econ Syst Res 21(3):187–216. https://doi.org/10.1080/09535310903541298
Niccolucci V, Bastianoni S, Tiezzi EBP, Wackernagel M, Marchettini N (2009) How deep is the footprint? A 3D representation. Ecol Model 220(20):2819–2823. https://doi.org/10.1016/j.ecolmodel.2009.07.018

Niccolucci V, Galli A, Reed A, Neri E, Wackernagel M, Bastianoni S (2011) Towards a 3D national ecological footprint geography. Ecol Model 222(16):2939–2944. https://doi.org/10.1016/j.ecolmodel.2011.04.020

O'Neill DW, Fanning AL, Lamb WF, Steinberger JK (2018) A good life for all within planetary boundaries. Nat Sustain 1:88–95. https://doi.org/10.1038/s41893-018-0021-4

Ostrom E (2009) A general framework for analyzing sustainability of social-ecological systems. Science 325(5939):419–422. https://doi.org/10.1126/science.1172133

Pearce DW, Atkinson GD (1993) Capital theory and the measurement of sustainable development: an indicator of "weak" sustainability. Ecol Econ 8:103–108. https://doi.org/10.1016/0921-8009(93)90039-9

Strezov V, Evans A, Evans TJ (2017) Assessment of the economic, social and environmental dimensions of the indicators for sustainable development. Sust Dev 25:242–253. https://doi.org/10.1002/sd.1649

Van den Bergh JCJM, Grazi F (2014) Ecological footprint policy? Land use as an environmental indicator. J Ind Ecol 18(1):10–19. https://doi.org/10.1111/jiec.12045

Van den Bergh JCJM, Verbruggen H (1999) Spatial sustainability, trade and indicators: an evaluation of the 'ecological footprint. Ecol Econ 29(1):61–72. https://doi.org/10.1016/S0921-8009(99)00032-4

Vasseur L, Horning D, Thornbush M, Cohen-Shacham E, Andrade A, Barrow E, Edwards SR, Wit P, Jones M (2017) Complex problems and unchallenged solutions: bringing ecosystem governance to the forefront of the UN sustainable development goals. Ambio 46(7):731–742. https://doi.org/10.1007/s13280-017-0918-6

Venetoulis J, Talberth J (2008) Refining the ecological footprint. Environ Dev Sustain 10(4):441–469. https://doi.org/10.1007/s10668-006-9074-z

Virapongse A, Brooks S, Metcalf EC, Zedalis M, Gosz J, Kliskey A, Alessa L (2016) A social-ecological systems approach for environmental management. J Environ Manag 178:83–91. https://doi.org/10.1016/j.jenvman.2016.02.028

Von Bertalanffy L (1950) The theory of open systems in physics and biology. Science 111(2872):23–29. https://www.jstor.org/stable/1676073

Wackernagel M (2014) Comment on Ecological Footprint policy? Land use as an environmental indicator. J Ind Ecol 18(1):20–23. https://doi.org/10.1111/jiec.12094

Wackernagel M, Beyers B (2019) Ecological footprint: managing our biocapacity budget. Global Footprint Network, Oakland

Wackernagel M, Kitzes J, Moran D, Goldfinger S, Thomas M (2006) The ecological footprint of cities and regions: comparing resource availability with resource demand. Environ Urban 18(1):103–112. https://doi.org/10.1177/0956247806063978

Chapter 3
Biocapacity Accounting

Abstract The focus of this chapter is on the biocapacity side of Ecological Footprint and biocapacity accounting. This is based on measuring natural capital for a land-based metric. Global hectares are the unit for establishing comparable portions of land around the world that support humanity's consumption. This represents the resource base (natural capital) that supports human life and activities. It can be characterised as 'carrying capacity' because it poses an environmental limit to human consumption. In this way, biocapacity is a fundamental part of the Ecological Footprint and biocapacity accounting. When the Ecological Footprint is set against biocapacity, a full picture of the environmental situation can be presented at different scales. Biocapacity acts like a 'measuring stick' for consumption and, thereby, performs a vital function in the accounts. This chapter addresses biocapacity as entailing production and, therefore, as a resource base for humanity. Biocapacity is what sets the Ecological Footprint apart from other performance indicators in that it provides a baseline for comparison, from where performance and sustainability can be assessed and monitored.

Keywords Carrying capacity · Natural capital · Resource base · Land-based metrics · Sustainability · Performance evaluation · Assessment and monitoring · Ecosystem services · Ecological overshoot

The focus of this chapter is on the biocapacity side of the Ecological Footprint and biocapacity accounting. This is based on measuring natural capital for a land-based metric. Global hectares are the unit for establishing comparable portions of land around the world that support humanity's consumption. This represents the resource base (natural capital) that supports human life and activities. It can be characterised as the 'carrying capacity' because it poses an environmental limit to human consumption. In this way, biocapacity is a fundamental part of the Ecological Footprint and biocapacity accounting. When the Ecological Footprint is set against biocapacity, a full picture of the environmental situation can be presented and interpreted at different scales (e.g., Wackernagel 2014). Biocapacity acts like a

M. J. Thornbush, *The Ecological Footprint as a Sustainability Metric*,
SpringerBriefs in Environmental Science,
https://doi.org/10.1007/978-3-030-62666-2_3

'measuring stick' for consumption and, thereby, performs a vital function in the accounts. This chapter addresses biocapacity as entailing production and, therefore, as a resource base for humanity. It is what sets it apart from other performance indicators in that it provides a baseline for comparison (Fang et al. 2015), from where performance and sustainability can be assessed and monitored.

For there to be resource appropriation, there first needs to be resources. Without biocapacity for comparison, it is difficult to gauge the status of the Ecological Footprint. For this reason, it is unrealistic to separate these two sides (EF and BC) of the accounts. Nevertheless, the ambition of this chapter is to consider biocapacity on its own in order to gain an understanding of it that is independent of its counterpart, the Ecological Footprint. Natural resources provide a basis for biocapacity as part of natural capital. Nature also provides ecosystem services that are often overlooked, but are a vital part of our natural inheritance and resource base.

This chapter focuses on biocapacity, but with the caveat that the Ecological Footprint is always hoovering in the background (as the two are inseparable and will only be isolated here for the purpose of scrutiny). The chapter begins by examining biocapacity from the perspective of carrying capacity, which is really addressing the environmental boundaries to consumption at the planetary level. It then adopts another angle by framing biocapacity through the lens of production – such as net primary production (NPP) – before considering the two sides of the Ecological Footprint and biocapacity when deliberating 'ecological overshoot' and our current status of global ecological deficit.

3.1 Biocapacity as 'Carrying Capacity'

How much natural capital is available at the planetary level or nationally operates as an upper limit to supply human demand. This defines the notion of 'carrying capacity' that is normally expressed at the global scale. It is what the Earth can sustain in terms of either human population (life) or the need for natural resources to supply food, energy, and so on, necessary for human activities.

It is not certain what the Earth's maximum capacity for the human population is, although it is expected that the world's population could reach as many as 12.7 billion people by 2100 (with 95% certainty, United Nations 2019a, p 5 – see Fig. 1). According to the United Nations Department of Economic and Social Affairs Population Dynamics (United Nations 2019a), it is projected that by 2030 the world's total population could reach 8.6 billion people (projected from data shown in Fig. 3.1 based on United Nations 2019b). However, the population growth rate has been diminishing (it is currently 1.1% per year, as in Fig. 3.2 based on United Nations 2019b), so this could be closer to 8.5 billion.

If we use Earth's existing biocapacity to indicate our current state of carrying capacity, it is possible that the Ecological Footprint and biocapacity accounting methodology can contribute to assessment and planning to prevent exceeding Earth's carrying capacity – in other words, living within Earth's planetary

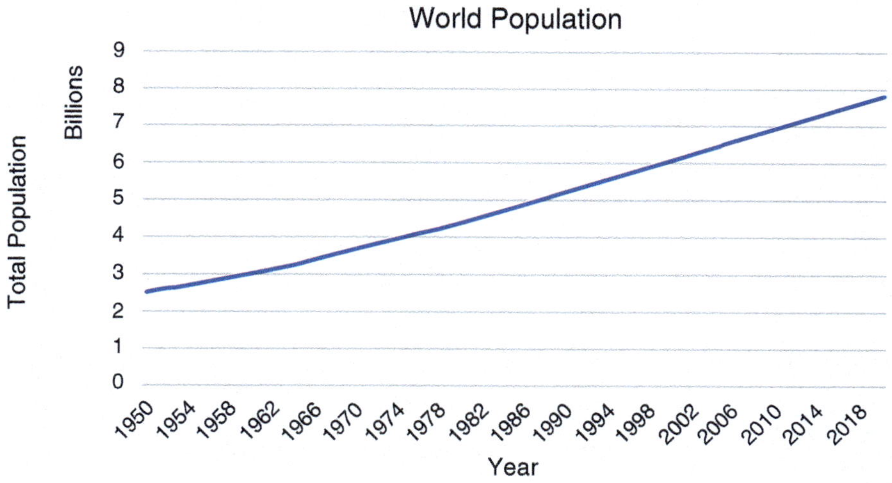

Fig. 3.1 Total world population, 1950–2020. (Source: United Nations 2019b)

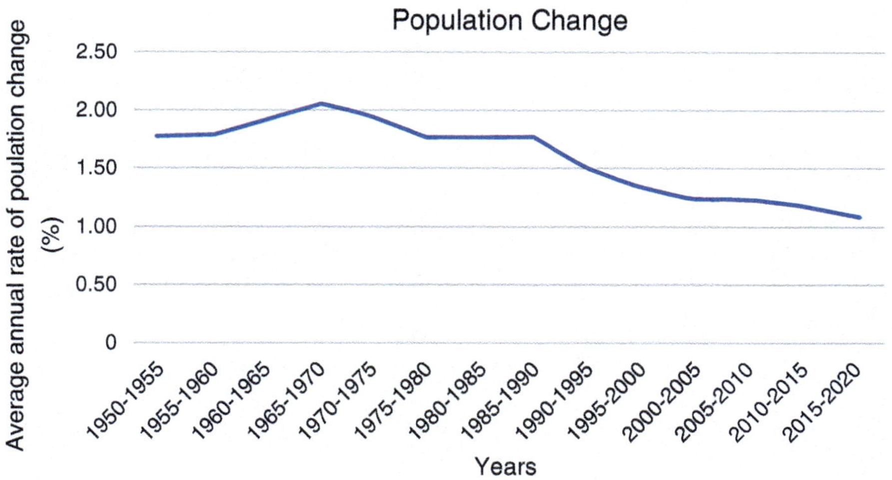

Fig. 3.2 World population change, 1950–2020. (Source: United Nations 2019b)

boundaries. There is some suggestion already, for instance, that the carrying capacity of the world was reached since 1970 – when both the Ecological Footprint and biocapacity were roughly 2.7 global hectares per person (Fig. 3.3, based on the Global Footprint Network 2020; see also Fig. 1 in Lin et al. 2018, p 7) – and has since been breached by our persistent demand for natural resources. These resources have been dwindling over the years (Fig. 3.4, based on data from the Global Footprint Network 2020), with most countries experiencing resource deficits.

If this is the case, then Earth's carrying capacity may be less than the upper limit used by the United Nations prediction models based on population growth. This

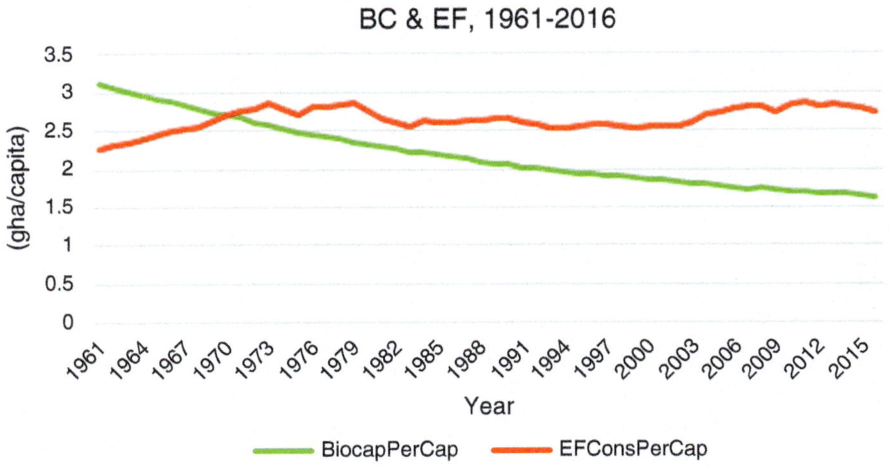

Fig. 3.3 The Ecological Footprint and biocapacity time series for the world, 1961–2016. (Source: Global Footprint Network 2020)

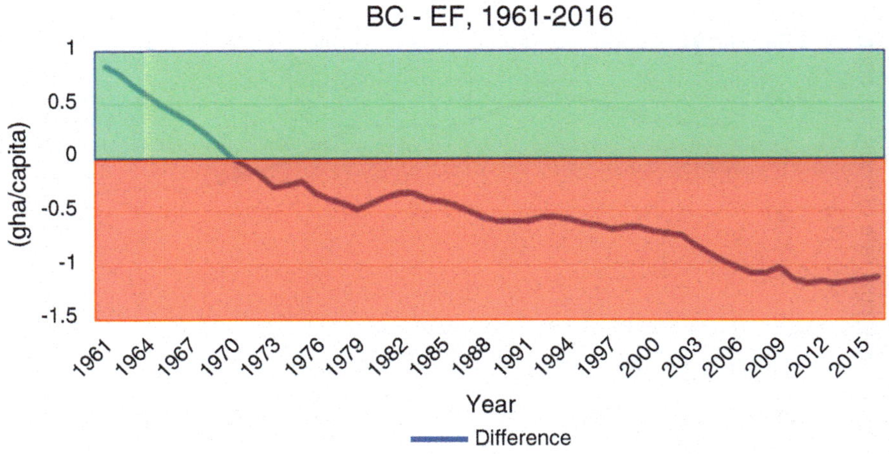

Fig. 3.4 The difference between world biocapacity and the Ecological Footprint, 1961–2016. (Source: Global Footprint Network 2020)

means that Earth's carrying capacity may be closer to 4 billion people for a population-based estimate (see Fig. 1 in United Nations 2019a, p 5). The implication of this is that we currently have a 'population overshoot' of almost 4 billion or nearly double Earth's carrying capacity. This overshoot has triggered resource depletion over time because of increased demand as the population continues to grow. This is especially true for the carbon Footprint (cF), as countries become more industrially developed and consumptive (Fig. 3.5, based on the Global Footprint Network 2020).

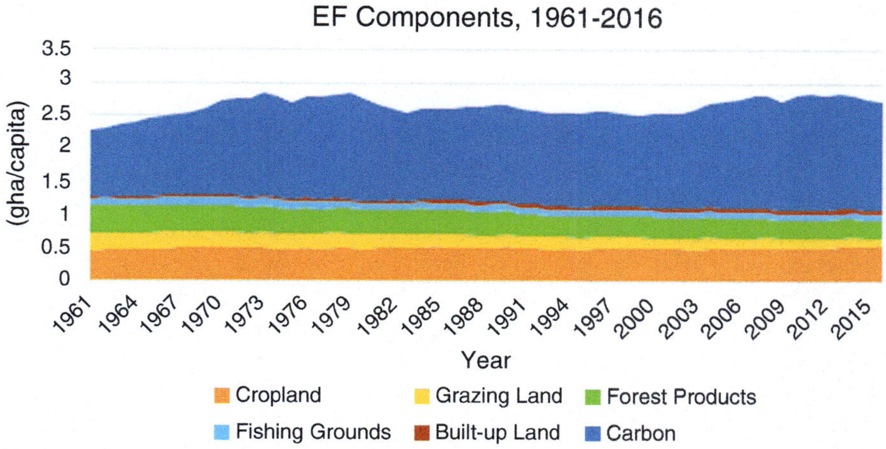

Fig. 3.5 Components of the world's Ecological Footprint, 1961–2016. (Source: Global Footprint Network 2020)

However, population (and human capital) alone cannot be the only basis for determining Earth's carrying capacity, and biocapacity (or natural capital) has an equally important role to play. Biocapacity – as the biologically productive area of the Earth's surface – can be measured using net primary production (NPP) based on biomass productivity. Satellite imagery can help us to discern this, as for example 31% of the Earth is covered in forest (Blomqvist et al. 2013; also see Giampietro and Saltelli 2014). All producers should be included, however – and not just forests. This includes crops (human production) among biocapacity.

3.2 Framing Production

In the balance between human and natural capital, humans operate like other animals as consumers. However, we are also considered to be producers because we affect production through crops and by planting trees and, thereby, so affect crop and forest lands. Grazing animals are also selected by humans as part of cultivation, so they too are part of our production. Fish farming is expanding these days and should likewise be included in human productivity calculations along with natural and artificial (e.g., fish farms) fishing grounds. Built-up land constitutes a growing area of human use and creation for city-dwelling and movement (transportation and infrastructure). These are the five components of biocapacity (production) that have counterparts in the Ecological Footprint (consumption). The latter includes crop and grazing lands used for food procurement as well as drawing upon fishing grounds for this consumption. Forests supply food products, but also wood products for use. Built-up land is used up for living and transport, when it could be available for other land uses, such as crop land (if suitable). Nevertheless, urban agriculture is growing and could construe a multiuse in urban areas.

The carbon Footprint is the only aspect of consumption that denotes a waste product based on the human release of CO_2 through respiration and is affected by human population size (and that of other animals). In addition, humans contribute to the production of this gas through combustion processes in energy production and so on – so that it is affected by manufacturing intensity and development level. Therefore, the carbon Footprint should reflect the latter (industrial development) more explicitly than the Ecological Footprint. Through acidification, it is also possible for humans to augment CO_2 production in soils, waters, and more, as carbonic acid develops in the presence of water. On the other hand, carbon is consumed through vegetation growth (of trees, but also other plants, including phytoplankton, crops, etc.). It is noteworthy to mention here that carbon sequestration is an ecosystem service. Soils capture carbon, as too do plants and animals, as an inorganic component of ecosystems. They are known to perform an essential ecosystem service of carbon capture and storage as a natural (physicochemical) carbon sink. Likewise, rocks (limestone) lock away carbon in the longer term and are also natural (physical) carbon reservoirs. Such storage is only temporary (either in the short- or longer term) and cannot represent carbon consumption. Therefore, it is necessary to decipher between flows and stocks of the components of this methodology.

3.3 Ecological Overshoot

Since the 1970s, the world has been in ecological overshoot where the Ecological Footprint exceeded biocapacity (see Fig. 3.3). This means that since then consumption (demand) has been greater than production (supply), causing ecological deficit in some countries, especially those with low natural capital. This discrepancy (between the Ecological Footprint and biocapacity) can present another way to quantify the severity of ecological deficit and, thereby, provide another measure of 'weak' versus 'strong' sustainability. The Ecological Footprint is known to be higher in developed countries due to the behaviour of such populations to consume goods and services rather than contributing towards production in agriculture and so on. According to Lin et al. (2018, p 9), the Ecological Footprint continues to grow annually at a rate of 2%. In a situation of global ecological overshoot since the 1970s, this suggests that the environment has been experiencing degradation in a situation of persistent and worsening global overshoot. However, the world's Ecological Footprint (measured in number of Earths) has stabilised in the past (e.g. early 1970s, 1980s, 1990s into 2000) and most recently seems to have plateaued since 2010 (Fig. 3.6, based on data from the Global Footprint Network 2020). Moreover, most recently the coronavirus pandemic (COVID-19) in 2020 is already known to have delayed Earth Overshoot Day by 3 weeks to August 22 (from July 29, 2019, refer to https://www.overshootday.org/).

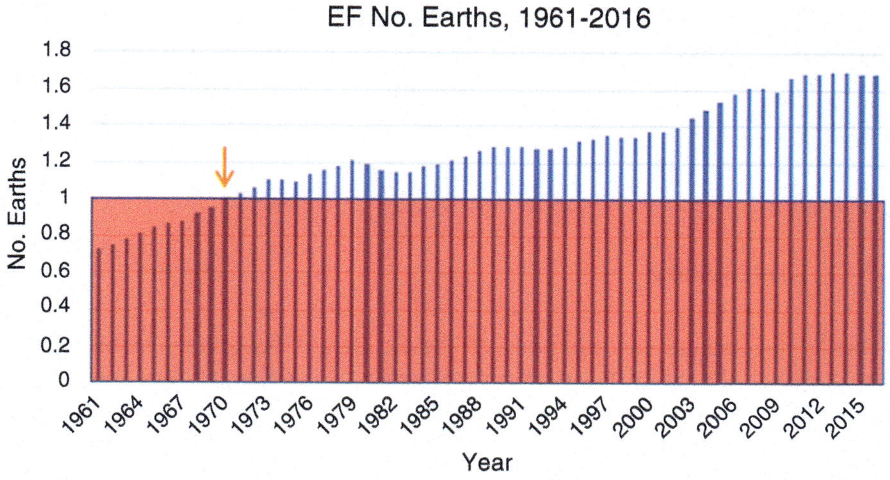

Fig. 3.6 The world's Ecological Footprint (bars), 1961–2016. (Source: Global Footprint Network 2020)

Table 3.1 Summary of a selection of national Footprint data

Location	Year	BC (gha/ capita)	BC (gha)	EF (gha/capita)	EF (gha)	EF (No. Earths)
Canada	2016	15.12	548788594.7	7.24	280880693.6	4.75
Canada	1970	23.12	496057008.2	9.22	197794068.4	3.41
China	2016	0.96	1,373,628,876	3.62	5,195,885,897	2.22
China	1970	0.91	765349868.3	1.06	890223488.4	0.39
Costa Rica	2016	1.55	7552943.64	2.68	13027739.12	1.65
Costa Rica	1970	4.3	7946907.51	2.51	4637842.85	0.93
USA	2016	3.65	1,174,980,442	8.1	2,611,053,058	4.97
USA	1970	4.6	964109444.8	10.37	2,172,630,366	3.83
World	2016	1.63	12,169,283,366	2.75	20,509,032,352	1.69
World	1970	2.71	10,016,193,932	2.72	10,073,811,315	1.01

Source: Global Footprint Network 2020

The implication of this is that more people need to live within the resources and ecosystem services provided by 1 Earth in order to prevent the Ecological Footprint from increasing over time (see Fig. 3 in Lin et al. 2018, p 10). Living within 1 Earth is equivalent to the biocapacity available on the planet, which is calculated to be 1.6 global hectares per person (for the Global Footprint Network 2020; 1.7 gha per person according to Lin et al. 2018, p 14 for the 2018 Edition of the National Footprint Accounts based on 2016 data). In this way, it is possible to sustain Earth's current population. However, even in some developing countries, the Ecological Footprint is greater than 1 Earth and has been so for the world as a whole since 1970 (see Fig. 3.6). According to Lin et al. (2018, p 15; also, Table 3.1), the current world

Ecological Footprint of 2.8 global hectares per person remains far above the biocapacity available to each person. In other words, the current world average of 1.7 Earths (based on the Global Footprint Network 2020) needs to be reduced either by curtailing consumption or increasing production (and/or efficiency) in order to counteract persistent overshoot.

3.4 Conclusions

One of the unique characteristics of the Ecological Footprint and biocapacity accounting is the ability to compare supply versus demand. In this way, it is possible to ascertain whether there is a net reserve or deficit. The latter is indicative of humanity overshooting its consumption beyond what is available from the Earth's natural resources. This chapter has focused on the balance used in the determination of the carrying capacity of an environment and to define production as well as ecological overshoot. Importantly, humanity needs to stay within the boundaries set by nature (e.g. 1 Earth) for sustainability to ensue. Admittedly, there is more to sustainability than just the environment, and it is necessary to also consider socio-economics as well as cultural aspects that may influence it.

In the next chapter, a case study from Costa Rica will provide the basis for analysis (Chap. 4) and discussion (Chap. 5) of a developing country. Rural peasant farmers (or '*campesinos*') located in the Alexander Skutch Biological Corridor will provide a basis for a cross-scalar examination of the Ecological Footprint. Importantly, by examining producers, it is possible to consider a more than typically 'balanced' population sample. The results will be compared to the national score for Costa Rica and deployed to scrutinise the human impact on the environment based on a developing country.

References

Blomqvist L, Brook BW, Ellis EC, Kareiva PM, Nordhaus T, Shellenberger M (2013) The ecological footprint remains a misleading metric of global sustainability. PLoS Biol 11(11): e1001702. https://doi.org/10.1371/journal.pbio.1001702

Fang K, Heijungs R, de Snoo GR (2015) Understanding the complementary linkages between environmental footprints and planetary boundaries in a footprint–boundary environmental sustainability assessment framework. Ecol Econ 114:218–226. https://doi.org/10.1016/j.ecolecon.2015.04.008

Giampietro M, Saltelli A (2014) Footworking in circles. Ecol Indic 46:260–263. https://doi.org/10.1016/j.ecolind.2014.06.019

Global Footprint Network (2020) National Footprint Accounts. https://data.footprintnetwork.org/#/

Lin D, Hanscom L, Murthy A, Galli A, Evans M, Neill E, Mancini MS, Martindill J, Medouar F-Z, Huang S, Wackernagel M (2018) Ecological footprint accounting for countries: updates and results of the National Footprint Accounts, 2012–2018. Resources 7(3):58. https://doi.org/10.3390/resources7030058

United Nations (2019a) World population prospects 2019: highlights. Department of Economic and Social Affairs Population Division, United Nations, New York, NY. https://population.un.org/wpp/Publications/Files/WPP2019_Highlights.pdf

United Nations (2019b) World population prospects 2019 (download files). Department of Economic and Social Affairs Population Dynamics. https://population.un.org/wpp/Download/Standard/Population/

Wackernagel M (2014) Comment on ecological footprint policy? Land use as an environmental indicator. J Ind Ecol 18(1):20–23. https://doi.org/10.1111/jiec.12094

Chapter 4
Case Study – Methods

Abstract This chapter employs a case study in an investigation of Costa Rica's national Ecological Footprint and biocapacity. Sampling farmers in the Alexander Skutch Biological Corridor as the study area, this chapter relays the methods used, including surveys based on the Footprint Calculator with follow-up interviews. In this way, it is possible to discern the regional environmental performance of these farmers and learn from them (at household–regional levels) how to improve their environmental performance. This contributes to an understanding of cross-scalar variations in the environmental performance of the Ecological Footprint and biocapacity. It also allows for a critical approach to the Footprint Calculator used to survey at these scales based on the online survey.

Keywords Peasant farmers · '*campesinos*' · Footprint Calculator · Farmers surveys · Household · Regional scale · Aggregation · Alexander Skutch Biological Corridor · Costa Rica

This chapter employs a case study in an investigation of Costa Rica's national Ecological Footprint and biocapacity. Sampling farmers in the Alexander Skutch Biological Corridor as the study area, this chapter relays the methods used in the study, including surveys based on the Footprint Calculator with follow-up interviews. In this way, it is possible to discern the regional environmental performance of these farmers and learn from them (at household–regional level) how to improve the environmental performance in a biological corridor. This contributes to an understanding of cross-scalar variations in environmental performance on the basis of the Ecological Footprint in particular. It also allows for a critical approach to the Footprint Calculator used to survey at these scales. Although the Footprint Calculator outputs information about consumption (and the Ecological Footprint), it is lacking details about production (biocapacity). Therefore, the function that biocapacity serves to contextualise the Ecological Footprint is missing and, thus, it is not possible to discern ecological overshoot and assess environmental sustainability to the degree possible with the National Footprint Accounts – at country level.

M. J. Thornbush, *The Ecological Footprint as a Sustainability Metric*,
SpringerBriefs in Environmental Science,
https://doi.org/10.1007/978-3-030-62666-2_4

Costa Rica was recently awarded as the United Nations Champion of the Earth for 2019. According to Delahaye (2019), in the policy leadership category especially, Costa Rica was recognised for its role in nature protection and fighting against climate change. The country is aiming to decarbonise, achieving net zero carbon emissions, by 2050 in accordance with the Paris Climate Agreement and the United Nations Sustainable Development Goals. To achieve this ambition, the country will not produce more emissions than can be offset through actions such as maintaining and expanding its forests. It already produces low emissions (0.4% of world emissions), but Costa Rica seeks to wholly produce renewable energy by 2030, when electric buses and taxis will be deployed. In 2017, also according to Delahaye (2019), the country ran on renewable energy for a record of 300 days; and 95% of its energy is already renewable. Costa Rica has counteracted deforestation since 1987 and has since become half-forested, which augments its biocapacity in order to offset emissions.

According to the National Footprint Accounts (Global Footprint Network 2020), in 2016 Costa Rica had an annual per person GDP of $9592 per person; with a total population of 4,857,274 people. Its biocapacity per person was 1.6 global hectares and its Ecological Footprint 2.7 global hectares per person (see Table 3.1), establishing a net deficit of −1.1 global hectares per person (based on the difference between the EF and BC, according to the Global Footprint Network 2020 – using data from 2016 in the 2020 Edition of the National Footprint Accounts). Figure 4.1 conveys the data since 1961, deploying the Ecological Footprint Explorer to output data for Costa Rica (country_code = 48) available online from the Global Footprint Network (2020) and published every year for over 200 countries and regions.

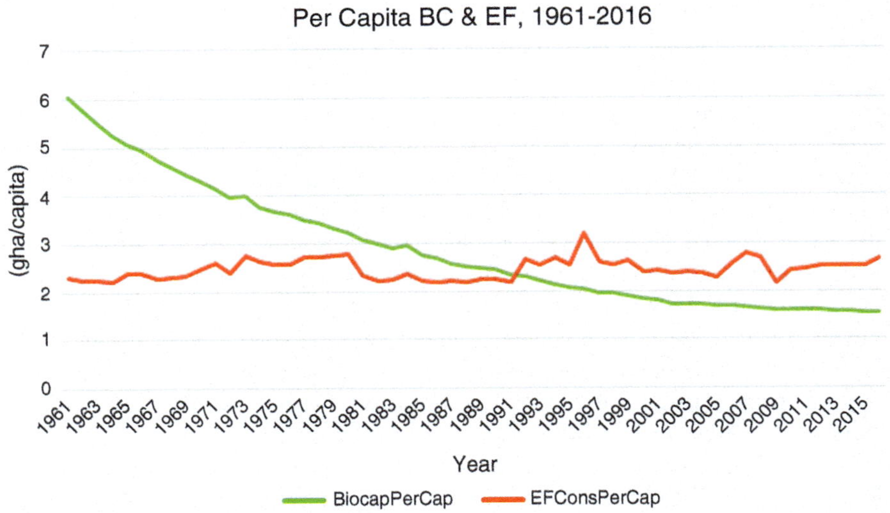

Fig. 4.1 Time series for Costa Rica's biocapacity and the Ecological Footprint, 1961–2016. (Source: Global Footprint Network 2020)

The data since 1961 convey a biocapacity that was initially around 6 global hectares per person and has since been gradually diminishing – to 1.6 global hectares per person in 2016 (see green line in Fig. 4.1). The Ecological Footprint was over 2 global hectares per person in 1961 and has since oscillated, but has not exceeded 3 global hectares per person (1996) – it reached 2.8 global hectares per person in 1979 and is currently approximately the same as before the 2008 financial crisis, being 2.7 global hectares per person in 2007 (see Fig. 4.1). This conveys some recovery in the country's economy that can be tracked through spending (consumption). The lines (of the Ecological Footprint and biocapacity) crossed in 1991 – a couple of decades later than 1970 for the world (as conveyed in previous chapters) – at 2.4 global hectares per person – see Fig. 4.1. As demonstrative of the level of consumption in the country, the Ecological Footprint is influenced by local demand, but is also sensitive to global events – such as the global financial crisis and most recently COVID-19. A reason for this is Costa Rica's link to global markets, as for instance in terms of food exports.

When the Ecological Footprint of consumption (EF_C) is subtracted from biocapacity (BC), as 'net' graphs (Fig. 4.2, based on 2016 data from the Global Footprint Network 2020), there is evidence of a steep drop in forest land that stabilised around 1986. There is also a decline evident since 1993 for grazing land and a steady decline in the fishing grounds component. Crop land remained stable

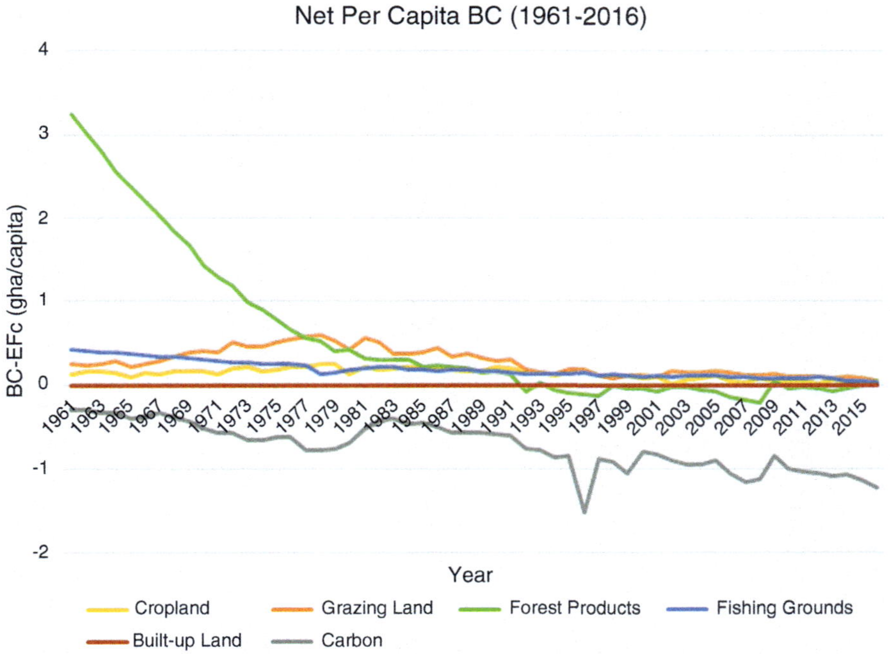

Fig. 4.2 Net per person biocapacity (BC – EF_C) for Costa Rica, 1961–2016. (Source: Global Footprint Network 2020)

between 1961 and 2016. Built-up land is problematic, as there is zero-change evident – which is questionable for the country. Also, the carbon Footprint is mostly represented in the EF_C because of the negative values conveyed by the carbon component. Actually, if one examines the country spreadsheet (Country_Trends), the column for the carbon component is zero for biocapacity (production). This is unrealistic because of the relationships evident between biocapacity and the carbon cycle, as for instance including carbon sequestration by plants. However, as noted in Chap. 1, a sequestration rate (constant) is deployed in the accounts to correct for this at the global level. Arguably, this represents the greatest challenge to precision in encapsulating actual biocapacity. It has been advocated (e.g., Hertwich 2011; Hertwich and Peters 2009) that household consumption is the greatest contributor to national carbon Footprints. Household income and householder age have been investigated among demographic variables as well as socioeconomic factors that influence the lifestyles of households (e.g., Jones and Kammen 2011; Shigetomi et al. 2014, 2015).

Since most countries are in ecological deficit (Rees and Wackernagel 2013), where their consumption exceeds their biocapacity (EF > BC), it is perhaps not surprising that Costa Rica has maintained a slight deficit, since most countries have. Wackernagel and Rees (1996) noted that a technological fix can have a major influence on sustainability through resource efficiency, allowing people not to have to reduce their material standards of living (as set in the framework by Haberl et al. 2004, Fig. 5, p 209). According to Wackernagel et al. (2006), ecological deficits can be reduced through various actions, including:

- by reducing the demand for natural capital with resource-efficient technology;
- reducing human consumption, but not necessarily the quality of life;
- reducing population size; and
- through investments in natural capital, such as by implementing methods of resource extraction that enhance bioproductivity.

These authors have espoused that, "...a Footprint smaller than the available biocapacity is a necessary condition for 'strong sustainability', a stance which asserts securing people's well-being" (p 108). We will now turn to examine this as it relates to the case study for Costa Rica.

4.1 Research Question

The research question driving this case study is:

- What is the Ecological Footprint of farming households located in the Alexander Skutch Biological Corridor of Costa Rica?

It was not possible to include an assessment of biocapacity in this investigation because of the use of the Footprint Calculator, which does not output biocapacity information. For this reason, as stated already, determining ecological overshoot at this scale was not possible. Scale-wise, the focus is on household–community level Footprints in the corridor, representing the local–regional spatial scale in southern Costa Rica. The location is the Alexander Skutch Biological Corridor, that connects the upland region to lower parts of the corridor along the main river valley. More specifically, Daugherty (2005) identified the central (core) location of the study area in the Alexander Skutch Biological Corridor as linking Los Cusingos Neotropical Bird Sanctuary with the Las Nubes Biological Reserve through the Río Peñas Blancas. This area is connected to Mexico and Panama as part of the Mesoamerican Biological Corridor.

4.2 Case Aim and Objectives

This research examines the Ecological Footprint at the household–community scale in order to decipher contributing variables affecting local–regional Footprints in the Alexander Skutch Biological Corridor, Costa Rica. Farmers located in this study area were approached to complete Footprint Calculator surveys to help inform their consumption across the main categories of the calculator, including food, housing, and transportation (http://www.footprintcalculator.org/). These categories were based on the Footprint Calculator online survey.

According to a news release by the Global Footprint Network (2010): "The calculator takes users through a series of questions about their lifestyle, and determines their Ecological Footprint – the amount of land area it takes to produce all the resources they consume and absorb their CO_2 emissions. It also identifies their greatest areas of resource consumption and translates this into the number of planets we would need if everyone lived that way." The Ecological Footprint can be deployed as an educational tool (Barrett et al. 2005; Lenzen and Murray 2003; McManus and Haughton 2006) to help communities understand the impacts that their lifestyle choices may have on the environment. In the current study, however, the reverse was the case – and the Ecological Footprint was informed by study participants who inadvertently provided lessons learned (refer to Sect. 5.2). As mentioned previously, this tool does not output information regarding biocapacity, so that it was not possible to determine how much of the biocapacity was being taken up by their Ecological Footprint. However, the national figures can help to inform this assessment. Furthermore, the 'one planet' approach (1 Earth) was adopted to replace biocapacity for comparisons in the sustainability assessment.

The Footprint Calculator has already been applied – with geographically-specific information – to countries such as the United States, Australia, Switzerland, and Calgary (Canada), including Spanish-speaking countries such as Ecuador, Peru, and Argentina (Global Footprint Network 2010). This is a relevant consideration because geographical location matters – in addition to lifestyle choices – in controlling national variables that may be impacting the Ecological Footprint due to societal influences on ecological demand (Global Footprint Network 2010). However, it should be noted here that this is a global calculator and does not differentiate between countries, using inputs from the National Footprint Accounts more broadly than country level. Consequently, it is necessary to adopt an integrated sociobiophysical approach, based on social-ecological systems – which, according to Haberl et al. (2004), have overlapping natural (biophysical) and cultural (symbolic) spheres of causation in order to calibrate the instrument to its location of use in interpreting the outputs.

The purpose or overarching aim of this case study was to assess the environmental performance of farmers in the study area in Costa Rica. More specifically, the objectives of this case study are:

1. To sample as broadly as possible within the study area in the Alexander Skutch Biological Corridor, Costa Rica.
2. To survey the farmers in the corridor using the Footprint Calculator to assess their Ecological Footprint (consumption only) in terms of food, housing, and transportation – which are the consumption categories in the Footprint Calculator.
3. To compare the regional findings with national results using the most recent available National Footprint Accounts.

This will allow the results of the study to inform one of the chief criticisms of the Ecological Footprint and biocapacity accounting regarding aggregation (see Chap. 6), as well as help to contextualise the findings for biocapacity at the national scale. Specifically, by examining a part of the country that is a biological corridor, it is possible to ascertain a lower boundary for the national results. From this, one can ask the question: What lessons can be learned from the corridor's Footprint? This will be the major contribution of using this case study in Costa Rica's corridor, in addition to a critical approach to using the Footprint Calculator.

4.3 Study Area Sampling

There are at least 2000 farmers located in the Alexander Skutch Biological Corridor, representing some 600 households (Luis Angel Rojas pers. comm 2020). Therefore, a sample of at least 96 participating farms was needed for an overall representation at the 95% confidence level (based on Eq. **4.1**). The minimum sample size was determined using the following equation:

$$n = \widehat{p}\,\widehat{q}\left(\frac{z\alpha/2}{E}\right)^2 \tag{4.1}$$

where $\widehat{p}$ ('*p* hat') is the sample proportion (*X*/*n*) and $\widehat{q}$ ('q hat') = 1 - $\widehat{p}$ or = *n* - *X*/*n* – where *X* = number of sample units that possess the characteristics of interest and *n* = sample size. $z_{\alpha/2}$ is based on the *z*-score and *E* is the margin of error based on the estimated accuracy. In this case, *E* is set at an error margin of 10%. The derived $\widehat{p}$ value is around 50% and $\widehat{q}$ is, therefore, set at 50%. This works out to be 600 samples

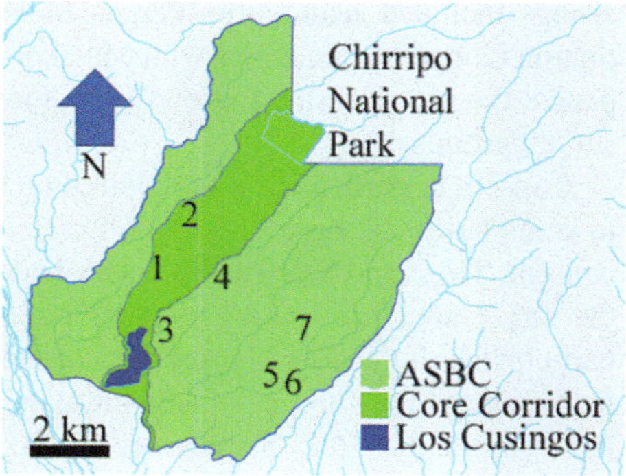

Fig. 4.3 The Alexander Skutch Biological Corridor – ASBC, including the towns: 1 = Santa Elena; 2 = Trinidad; 3 = Quizarrá; 4 = Montecarlo; 5 = San Francisco; 6 = Santa Marta; and 7 = San Ignacio (Source: https://cobas.juturna.ca/#/)

required based on a two-tailed distribution set to cover a broader range of the distribution due to uncertainty, where $z_{\alpha/2} = 1.96$. The required minimum sample size is, therefore 96 households.

Based on Fig. 4.3 (modified from Monge 2018, p 38, Fig. 3), it is possible to estimate the land area in the core corridor (in bright green), which spans elevations between 500 and 2500 m located in the greater corridor (ASBC in green). Towns in the study area are situated between 1500 and 2000 m in elevation (see Fig. 4.3).

The study area towns presented in this study include Santa Elena (SE, 1, including the '*barrio*' or neighbourhood of Trinidad, 2), Quizarrá (QA, 3), Montecarlo (MO, 4), San Francisco (SF, 5, and Santa Marta, 6), and San Ignacio (SI, 7). The latter three towns (or '*pueblos*') were added because of a limited sample size in Montecarlo. These towns allowed for broader sampling from the interior (core) corridor outwards towards its eastern border, roughly dissecting the corridor across the middle (see Fig. 4.3). Survey areas were chosen based on landmarks located throughout the towns and systematically sampled along the roadways in vicinity of these landmarks (e.g., Coffee Pan, library, football field, school, etc. in Santa Elena), which usually entailed main roadways. Snowballing was integrated by exclusively following up on 'producing' households, where there were farmers or those with crops or livestock, as indicated by participating neighbours. Also, our driver and field guide (Alex Fonseca) was instrumental in directing us to people that he knew were *campesino* households.

The execution of the survey-interviews was performed in Spanish-speaking groups in February to March 2020 – after the Faculty of Environmental Studies Research Committee at York University approved the research proposal. Surveying was supervised by the lead researcher (the author), who trained surveyors before initiating the fieldwork. This involved going over the paperwork to be completed during the course of the fieldwork, as outlined in the next section. Importantly, our driver and field guide (Alex Fonseca) became the first participant in the study, so that the fieldworkers could practise executing the work in a safe environment before heading out. Each group comprised a female and male because women – who were at home during the day, when the survey-interviews were executed – felt more comfortable when answering the door and seeing another female. For this reason, the groups included Jean Carlo Vargas Mena accompanied by Mary Thornbush and Sharon González Benavides with Maickel Benavidas Jiménez. Alex Fonseca accompanied the group during these visits to monitor and further oversee the safety of the survey groups.

Consent to participate was acquired on the spot by the surveyors (without signing in some cases) because of a known preference by corridor participants to give oral consent. This approach was necessary because of a known lack of trust by farmers in the study area for signing to give consent. However, in practice, some of the participants did end up signing the consent form because they knew our driver and field guide (Alex Fonseca) – who ultimately functioned as 'our gatekeeper' to access the communities (cf. Spencer et al. 2014). Incentives were not systematically deployed in this study, and we relied on the goodwill of *campesinos* to participate. However, wherever possible, we offered to purchase something of interest from participants as a type of incentive, including plants, and more. Participants were very kind and, in some cases, offered for us to sample some of their produce, such as organic oranges, and some offered us refreshments. Because this is a well-researched area, there was an added question at the end of the survey (in the follow-up questions part) asking participants how many times prior they had been asked to participate in such research.

4.4 Data Collection

The study incorporated both quantitative and qualitative research methods in a mixed methods approach (after Gray and Morant 2003). This was executed using survey-interviews comprising both closed-ended questions (in the Footprint Calculator survey portion) and open-ended questions in a follow-up short interview with participants. The latter was designed to accommodate up to 5 minutes of interaction with participants in order to minimise the overall time of the survey-interviews. The research process involved five stages (Box 4.1) executed by a surveying team of three *campesinos* from the study area accompanied by the author as lead researcher, in two teams of two surveyors, and took no more than a half-hour (and, ideally, as close as possible to 20 minutes) at each household. All five stages are elaborated in the next subsections.

Box 4.1 Stages in the Research Process

Stage 1: Informed Consent (2–5 min) – conveying information about the research topic and main study objectives; oral consent (signed form by researcher on behalf of the participants) – if agreed, proceed to Stage 2.

(continued)

Box 4.1 (continued)

Stage 2: Demographic Information (2 minutes) – written responses to preliminary questions regarding demographics and farm attributes – proceed to Stage 3 when completed.

Stage 3: Farmer Survey (10–15 min) – closed-ended questions, with responses written on paper and later transferred to online Footprint Calculator (via the Global Footprint Network website: http://www.footprintcalculator.org/) to obtain a final score – proceed to Stage 4 when completed.

Stage 4: Follow-Up Questions (5 min) – open-ended questions structured around responses to Stage 3 survey; any added commentary – proceed to final Stage 5 when completed.

Stage 5: Wrap-Up (1–3 min) – any final questions or clarification on any of the previous stages.

Each of the two groups followed the protocol outlined in Appendix 4.1 based on these five stages. The datasheet was completed by the same people in each group during the fieldwork. It was designed to accommodate Spanish as the spoken language, so that all questions in the Footprint Calculator survey were translated using the Footprint Calculator website (https://www.footprintcalculator.org/) *Español* version.

4.4.1 (1) Informed Consent

Because of the aforementioned known distrust among *campesinos* to give signed informed consent, this was attained orally on site. In the first step of the research process, informed consent was obtained at each household when potential participants were first approached. Surveyors briefly introduced the research scope and ambition. If those producing households approached agreed to participate, then surveyors were instructed to sign on their behalf, denoting oral consent. In some cases, however, participants agreed to sign the consent form. Several participants expressed interest in having follow-up information sent to them in some way (e.g. pamphlet or presentation) to disseminate the study findings and inform them of the study findings. Their addresses were noted in anticipation of this future contact. Importantly, those approached were household heads who represented the entire household.

4.4.2 (2) Demographics

The purpose of collecting demographic information was to know more about the study participants in order to have details of their household configuration and farms independently of the Footprint Calculator survey and follow-up questions. Because the question asking about the number of people in their households was already included as part of the Footprint Calculator survey (see Question 5 in Appendix 1.1), this was excluded from the demographic questions. These were closed-ended questions (with set ranges and options). Annual income was asked in US dollars rather than in the national currency of colones. The reason for this was because the options for this question were based on a GDP per person of 9592 for 2016 (Global Footprint Network 2020, Open Data Platform), entered in the survey as US$10,000. However, there was a mixture of currency responses noted (see Appendix 4.1).

4.4.3 (3) Farmers' Surveys

The Footprint Calculator is available online from the Global Footprint Network (https://www.footprintcalculator.org/). It uses three major consumption categories (food, housing, and transportation) to determine the Ecological Footprint of those completing the survey – at whichever scale deployed. In this case, households of peasant farmers '*campesinos*' form the level of analysis, which was aggregated locally to encompass the towns in the study area within the corridor. Sample size for the farmers' surveys was based on a minimum of 96 surveys (see Sect. 4.3). To reiterate from Sect. 4.3, this was established based on a margin of error set at 10% and equal probabilities (of previously unknown standard deviation) set to 50% for a two-tailed distribution at 5% significance level ($p = 0.05$, 95% confidence level).

The Footprint Calculator survey is available in Spanish from the Global Footprint Network website. A summary of the questions asked appears in Appendix 1.1 (see Appendix 4.1 for the Spanish version). These questions are used in combination with the National Footprint Accounts to determine the global hectares consumed for each surveyed household. In practice, however, not all these questions were asked in order to keep the time of the survey-interviews closer to 20 minutes. For this reason, questions that participants would not have known the answers to (e.g., Q12 on car efficiency, Q9 on renewable energy) or irrelevant questions to the study area (e.g., Q13 on carpooling, Q15 on flying) were excluded and 'general' responses used, as indicated (in underlined, bold) in Appendix 1.1. Moreover, some answers tended to be repeated (e.g., Q3 on housing type, Q4 on house construction – to some extent, as these were the same across households in the study area). In fact, some became so obvious that it was unnecessary to ask them, and they were dropped later in the study to reduce the survey time; for example, all properties had electricity (Q7) and nobody travelled by train (Q14) – as there is no local train service. This will be discussed in detail later in the next chapter (Chap. 5) as part of the fieldwork findings.

According to David Lin (pers comm 2020) of the Global Footprint Network (Oakland, CA), there are quite a few steps before even beginning the calculation used by the Footprint Calculator. The National Footprint Accounts is the base dataset, which must be categorised into consumption categories. This is done first by applying the National Footprint Accounts as an environmental extension to a Multi-Region-Input-Output (MRIO) model. From this MRIO model's outputs, the Global Footprint Network then develops the Consumption Land-Use Matrix (CLUM) based on category concordances. Then CLUM serves as the basis of a hybrid top-down and bottom-up calculator. All these processes are quite complex, and there are very few publications on this. One recent publication by Collins et al. (2020) addressed role of personal Footprint calculators in helping people understand how to live within the limits of one planet. This article also contains a mini review of previously published articles and books dealing with calculators to consider the contribution of personal Footprint calculators.

4.4.4 (4) Follow-up Interviews

Semi-structured follow-up interviews were based on the following main survey questions:

1. What type of crops do you grow, and in what proportion?
2. How much do you grow in a year?
3. Where do you sell your crops?
4. Are you participating in Fair Trade or organic farming?
5. Do you use fertilisers or pesticides?
6. Do you benefit from Payments for Ecosystem Services?
7. Have you ever been approached before to complete a survey or an interview?

These questions were administered in person in the presence of a translator, obtained most often with oral (rather than written) informed consent. Up to four survey-interviews per group per day across 12 days in February to March 2020 was enough to draw a sufficiently large sample in excess of 96 farming households located in the corridor to enable engaging with a variety of household demographics, farm sizes, and crop types. The interviews were structured around the survey responses from each farmer survey (completed by household heads) based on the Footprint Calculator, divided into the three main sections of the consumption categories: food, housing, and transportation. The short follow-up interviews were intended to supplement the closed-ended questions on the farmer survey (based on Footprint Calculator), and expanded upon these to provide context using a few open-ended questions and any added commentary.

4.4.5 (5) Wrap-Up

In this final portion of the research process during the fieldwork component, participants were asked for questions of their own either to clarify the research process, tools or instruments, and the study's rationale. They had the opportunity to clarify any confusion or voice any concern. They could also add to their answers at this point or return to a question or point. Importantly, a final question is whether they had been previously approached (at any time before) to volunteer as research participants in other studies. The reason for this question was to assess whether individual participants, as well as the survey sample, has been overexposed to studies and, consequently, is considered to be 'exhausted'.

4.5 Data Analysis

The quantitative results based on the Footprint Calculator (farmers' surveys) contained details regarding Earth Overshoot Day – the date on which as much from nature has been extracted as the Earth can renew the entire year – e.g. July 29, 2019 (https://www.overshootday.org) for all of humanity. It was August 10, 2019 for Costa Rica (refer to Sect. 2.1). Generally, Earth Overshoot Day has been arriving earlier each year for the world tied to the number of Earths that humanity requires to support its resource needs and lifestyles (see Fig. 3.6). However, because of the COVID-19 epidemic, this has been reduced to August 22, 2020 most recently for the world (https://www.footprintnetwork.org/our-work/earth-overshoot-day/). The Global Footprint Network is asking that such a contraction of the Ecological Footprint occur by design rather than disaster.

A vital output for data analysis is the Ecological Footprint itself in global hectares (gha) – defined as the biologically productive (or 'bioproductive') area with world average productivity that is required to provide what the person consumes. Currently, 1.75 Earths are needed to sustain human consumption (Global Footprint Network 2020), which is determined by dividing all the planet's productive area by the number of its inhabitants. This result can be compared at different scales to the entire planet, at national level, regionally, and so on – so that a regional comparison can be made within the study area. By adopting the Footprint Calculator, it is possible to discern individual–household responses that can be aggregated to the household–community level (e.g., towns) in the determination of a local–regional Ecological Footprint (local representing the towns and regional the Alexander Skutch Biological Corridor).

The qualitative analysis of follow-up (semi-structured) interview content was transcribed in English onto a summary table (in addition to other tables available for demographics and the results of the Footprint Calculator, which are available in a spreadsheet). These results will appear in the next chapter (Chap. 5). This content was linked to household demographic information attained from participating farms.

The entire process of documenting the follow-up interviews involved an analytic journey (cf. Spencer et al. 2014) based on four stages:

1. **design and sampling** – where the Footprint Calculator survey was transcribed from the Global Footprint Network website (https://www.footprintcalculator.org/) and the study area and towns chosen based on preliminary visits to the corridor and maps (e.g., Montoya and Martínez 2015, p 17), as well as a calculated lower limit for sample size to be as evenly as possible (broadly) distributed among the corridor towns;
2. **data generation/collection** – during the fieldwork phase, through the execution of the survey-interviews in Spanish-speaking groups, where responses were recorded on paper by completing the fieldwork datasheet (see Appendix **4.1**) after informed consent was attained (usually orally, but sometimes in writing);
3. **analysis** – that involved amassing statistics and creating tables and graphs for the dissemination of the results to be deployed in the final stage; and
4. **reporting** – covered in the next chapter, denoting the research findings.

Appendix 4.1: Fieldwork Datasheet

(2) Demographics: (2 min) Participant age and gender	Age: <30/30–60/>60 years old Gender: Male/Female
Are you an owner or tenant farmer?	Owner/tenant
Years at farm	<1/1–10/>10 years
Farm size	<10/10–20/>20 hectares
Annual income	<US$10,000/US$10–40,000/>US$40,000
(3) footprint calculator: (15 minutes) **ALIMENTOS** 1. Con qué frecuencia comes productos pecuarios?	Nunca/Ocasionalemente/Frecuente/Muy Frecuente (carne roja, cerdo, pollo, pescado, huevos, productos lácteos)
- carne de vaca o cordero?	___%
- Cerdo?	___%
- Pollo?	___%
- Pescado o mariscos?	___%
- Huevos, queso y/o productos lácteos?	___%
2. Cuántos de los alimentos que comes son no procesados, no envasados o cultivados localmente?	Ningunos ___% Todos (menos que 320 kilómetros = km/200 milles de distancia)
- Qué cantidad de tu dieta se compone dealimentos frescos, sin embalaje?	___%
- Cuánto de tu dieta es cultivada or producida localmente?	___% (a menos de 320 km/200 millas de distancia)
ALOJAMIENTO 3. Qué tipo de vivienda describe tu hogar?	Independiente, sin agua corriente/ Independiente, con agua corriente/ Departamento de varios pisos/duplex, casa adosada, o edificio con 2–4 unidades/ Condominio de lujo

(continued)

4. Con qué material está construida tu casa?	Paja o bamboo/Madera/Ladrillo o cemento/ adobe/Acero o otro
5. Cuántas personas viven en tu hogar?	Solo yo (1)/... 10+
- Cuál es el tamaño de tu casa?	Pequeñtio (5 m^2 o 50 pies cuadrados)/Pequeño 16 m^2 o 175 pies cuadrados/Mediano 49 m^2 o 523 pies cuadrados/Grande 141 m^2 o 1520 pies cuadrados/Gigante 465 m^2 o 5008 pies cuadrados
6. Tienes electricidad?	No/Si
- Qué tan eficiente es tu casa en el consumo de energía?	Poco = Muy ineficiente/por debajo del promedio/Promedio/por encima del promedio/ Diseño centrado en la eficiencia = Muy
7. Qué porcentaje de la electricidad de tu casa proviene de Fuentes renovables?	Bajo ___% alto (directamente o mediante la compra de energía limpia)
8. Comparado con tus Vecinos, cuánta basura generas?	Mucho menos/Menos/Igual/Más/mucho más
- Qué es lo más parecido a tus compras mensuales de ropa nueva, calzado y/o artículos deportivos?	Mínimo a ninguno/no mucho/Promedio/por encima del promedio/mucho
- con qué frecuencia compras una nueva? ELECTRODOMÉSTICOS	Nunca, raramente/Infrecuentemente/ Ocasionalmente/A menudo/Muy a menudo
ELECTRÓNICA Y GADGETS	Nunca, raramente/Infrecuentemente/ Ocasionalmente/A menudo/Muy a menudo
LIBROS, REVISTAS Y PERIÓDICOS	Nunca, raramente/Infrecuentemente/ Ocasionalmente/A menudo/Muy a menudo
- Qué se acerca más a sus compras anuales de muebles Para el hogar?	Mínimo a ninguno/no mucho/Promedio/por encima del promedio/mucho
- Cuánto reciclas? PAPEL	Poco o nada/Algunos/Mitad/Más/Todo
EL PLASTICO	Poco o nada/Algunos/Mitad/Más/Todo
MOVILIDAD 9. Qué tan lejos viajas en automóvil o motocicleta cada semana? (Como conductor o pasajero)	- Automóvil: Cero (0 km o 0 milles) ... Muy Legos (800 km o 500 milles) - Ciclomotor: Cero (0 km o 0 milles) ... Muy Legos (800 km o 500 milles)
10. Qué distancia viajas en transporte público cada semana? (autobús, tren, etc.)	No tan lejos (0 km or 0 milles) ... Muy lejos (800 km o 500 milles)
- Qué distancia viajas cada semana por tren?	0 km o 0 milles ... 800 km o 500 milles
- Qué distancia viajas cada semana por autobús?	0 km o 0 milles ... 800 km o 500 milles
11. Qué distancia viajas por avión cada año?	Nada (0 horas) ... mucho (200 horas)
Results:	Earth overshoot day: No. earths: Component: Crop land ___/grazing land ___/Forest land ___/fishing grounds ___/built-up land ___/carbon ___ Category: Food ___/Housing ___/mobility ___/goods ___/services ___ EF: cF:

(continued)

(4) follow-up questions: (5 minutes) 1. What type of crops do you grow, and in what proportion? 2. How much do you grow in a year? 3. Where do you sell your crops? 4. Are you participating in fair Trade or organic farming? - do you use fertilisers or pesticides? 5. Do you benefit from payments for ecosystem services?	
(5) wrap-up (3 min) - have you ever been approached before to complete a survey or an interview? If so, how many times and when? - do you have any questions or comments to add?	

References

Barrett J, Birch R, Cherrett N, Wiedmann T (2005) Exploring the application of the ecological footprint to sustainable consumption policy. J Environ Policy Plan 7(4):303–316. https://doi.org/10.1080/15239080500441095

Collins A, Galli A, Hipwood T, Murthy A (2020) Living within a one planet reality: the contribution of personal footprint calculators. Environ Res Lett 15(2):025008. https://doi.org/10.1088/1748-9326/ab5f96

Daugherty HE (2005) Biodiversity conservation and rural sustainability: a case study of the Alexander Skutch biological corridor in southern Costa Rica. WIT Trans Ecol Environ 81:155–161. https://doi.org/10.2495/ECO050171

Delahaye J (2019) Costa Rica named UN Champion of the Earth in fight against climate change. Mirror (24 September 2019). https://www.mirror.co.uk/travel/central-south-america/costa-rica-named-un-champion-20166387

Global Footprint Network (2010) Earth Day Network and Global Footprint Network unveil new Footprint Calculator. http://www.footprintcalculator.org/

Global Footprint Network (2020) National Footprint Accounts. https://data.footprintnetwork.org/#/

Gray LC, Morant P (2003) Reconciling indigenous knowledge with scientific assessment of soil fertility changes in southwestern Burkina Faso. Geoderma 111(3–4):425–437. https://doi.org/10.1016/S0016-7061(02)00275-6

Haberl H, Fischer-Kowalski M, Krausmann F, Weisz H, Winiwarter V (2004) Progress towards sustainability? What the conceptual framework of material and energy flow accounting (MEFA) can offer. Land Use Policy 21:199–213. https://doi.org/10.1016/j.landusepol.2003.10.013

Hertwich EG (2011) The life cycle environmental impacts of consumption. Econ Syst Res 23(1):27–47. https://doi.org/10.1080/09535314.2010.536905

Hertwich EG, Peters GP (2009) Carbon footprint of nations: a global, trade-linked analysis. Environ Sci Technol 43:6414–6420. https://doi.org/10.1021/es803496a

Jones CM, Kammen DM (2011) Quantifying carbon footprint reduction opportunities for US households and communities. Environ Sci Technol 45:4088–4095. https://doi.org/10.1021/es102221h

Lenzen M, Murray SA (2003) Ecological footprint – issues and trends. Ecol Footprint:27

McManus P, Haughton G (2006) Planning with ecological footprints: a sympathetic critique of theory and practice. Environ Urban 18(1):113–127. https://doi.org/10.1177/0956247806063963

Monge AJ (2018) Environmental problem-solving: an application of institutional theory and systems thinking to the Alexander Skutch Biological Corridor. http://hdl.handle.net/10315/36027

Montoya F, Martínez AM (2015) Birds of the Alexander Skutch biological corridor, Costa Rica. Faculty of Environmental Studies, Las Nubes Project, York University, Toronto

Rees WE, Wackernagel M (2013) The shoe fits, but the footprint is larger than earth. PLoS Biol 11 (11):1–3. https://doi.org/10.1371/journal.pbio.1001701

Shigetomi Y, Nansai K, Kagawa S, Tohno S (2014) Changes in the carbon footprint of Japanese households in an aging society. Environ Sci Technol 48(11):6069–6080. https://doi.org/10.1021/es404939d

Shigetomi Y, Nansai K, Kagawa S, Tohno S (2015) Trends in Japanese households' critical-metals material footprints. Ecol Econ 119:118–126. https://doi.org/10.1016/j.ecolecon.2015.08.010

Spencer L, Ritchie J, Ormston R, O'Connor W, Barnard M (2014) Analysis: principles and processes. In: Ritchie J, Lewis J, Nicholls CM, Ormston R (eds) Qualitative research practice: a guide for social science students and researchers. Sage, Washington, DC

Wackernagel M, Kitzes J, Moran D, Goldfinger S, Thomas M (2006) The ecological footprint of cities and regions: comparing resource availability with resource demand. Environ Urban 18(1):103–112. https://doi.org/10.1177/0956247806063978

Chapter 5
Case Study – Findings

Abstract In this chapter, the results of the case study are presented. This chapter considers the findings in light of the demographics of study participants and conveys the follow-up interview data in addition to the results generated by the Footprint Calculator. The findings reveal an Ecological Footprint of nearly 3.0 global hectares, that is slightly greater than the national score of 2.7 global hectares based on the 2020 Edition of the Global Footprint Network's National Footprint Accounts. Consequently, the average Earth Overshoot Day was similar to the national result, especially for 2020. There were very few cases where people in the corridor were living within 1 Earth. On average, their lifestyle required closer to 2 Earths. Income was found to covary strongly in a linear relationship with both the Ecological and carbon Footprints. The latter (cF) and crop land were the most critical components, although grazing land and fishing grounds showed the most variance. The consumption category with the highest average was food – it also had the highest standard deviation among these categories. The main reasons for this will be taken up in the next chapter (Chap. 6), along with the solutions from the outputs of the Footprint Calculator. The towns in the corridor scored similarly on the Footprint Calculator, although there were sizeable standard deviations in the responses.

Keywords Footprint Calculator · Ecological Footprint · Number of Earths · Global hectares · Footprint components · Land types · carbon Footprint · Consumption categories · Earth Overshoot Day

It should be noted at the onset of this chapter that the participation rate in this study was extremely high. Very few households declined being part of the study because they had either already participated in a study (e.g., the previous day) and did not wish to partake in another survey or for no given reason. This makes for a participation rate exceeding 95% (for a total of 120 participating households). Only producing households were included in this sample, so some of them could not participate in the study, although they would otherwise have done so (and these were excluded from the participation rate). People were very generous with their time,

M. J. Thornbush, *The Ecological Footprint as a Sustainability Metric*, SpringerBriefs in Environmental Science,
https://doi.org/10.1007/978-3-030-62666-2_5

especially as they were not normally compensated, and answered all questions to the best of their ability and understanding. However, they were more likely to participate if the survey-interviews did not exceed 20 minutes.

We will now turn to the findings of the case study based in the Alexander Skutch Biological Corridor of Costa Rica. These results are presented here according to the general structure of the survey-interviews, according to demographics, farmers' surveys, and follow-up interviews. Some discussion appears alongside these findings for elaboration and to provide some context.

5.1 Findings with Discussion

There are 14 results (in gha) among the Footprint Calculator outputs. Additionally, there are another two results – one is the identification of Earth Overshoot Day and, the other, the proportion of the carbon Footprint set against the total Ecological Footprint. Besides getting an Ecological Footprint calculation in global hectares, one is also provided in the number of Earths (no. Earths) required to sustain the lifestyle of the person who completed the survey as representative of their household. Again, although there is no biocapacity for comparing the Ecological Footprint outputs, the number of Earths can be deployed for comparison to a one planet (1 Earth) level needed to sustain resources at the global scale. As noted in Box 5.1, to reiterate, it is ideal to keep the latter to 1 Earth in order to prevent ecological overshoot at a global scale. The results are also conveyed using land types (for the six components) as well as five consumption categories, including food, housing, and transportation or mobility as well as goods and services that are also outputted with the results. These results are output for each participant (household heads representing household consumption) in the survey-interviews, with the results presented in this chapter according to the towns sampled as well as overall for the corridor.

Box 5.1 Footprint Calculator Outputs

Your personal Earth Overshoot Day is: E.g., 16 May.

[Information: If everyone lived like you, Earth Overshoot Day would be on this day. This means by this date we would have used as much from nature as Earth can renew the entire year. In 2018, Earth Overshoot Day for humanity is August 1.]

If everyone lived like you, we would need E.g., 2.7 Earths

[Information: How do you compare? Average number of Earths: USA 5, Germany 3.2, China 2.1, S. Africa 2.0, and Brazil 1.8].

Why can't I get my Footprint score within the means of one planet?

(continued)

Box 5.1 (continued)

By land type – Built-up land, Forest products, Crop land, Grazing land, Fishing grounds, carbon Footprint.

By consumption category – Food, Shelter (or Housing), Mobility, Goods, Services.

E.g., *4.6 Your Ecological Footprint (global hectares or gha).

[Information: Your Ecological Footprint is the biologically productive area required to provide everything you consume. The Ecological Footprint can be compared to biocapacity, which is the productive area that exists on our planet, in your country, or in your region.]

A global hectare (gha) is a biologically productive hectare with world average productivity. A hectare is 10,000 square metres or about 2.5 acres. Currently, our planet has about 1.7 global hectares available per person. That's all the Earth's productive area divided by the number of people alive today.]

E.g., *7 Your carbon Footprint (CO_2 emissions in tonnes per year).

E.g., *53 Your carbon Footprint (% of your total Ecological Footprint).

Transcribed by Mary Thornbush, Faculty of Environmental Studies, York University (Email: mthornbu@yorku.ca) based on: http://www.footprintcalculator.org/.

5.2 Demographics

Table 5.1 provides a summary of the descriptive statistics (average ± standard deviation, sample size) for the different locations sampled during the course of the fieldwork. Across the six towns, overall, in the corridor sample there were 36% female participants in this study. Participants' average age was 55 ± 15 years (n = 117) and they had been at their farms an average of 26 ± 18 years (n = 111). Their average farm size was 6 ± 13 hectares (n = 114) and the income derived from that property was US$1641 ± $2294 based on those participants that revealed their specific income (n = 55) – rather than the set ranges. San Ignacio represented the youngest sample on average. Santa Marta had participants who were the longest (on average) on their farms, although the sample size was very small in these towns. The average declared farm size appeared to be largest in San Francisco, and smallest in Montecarlo and Santa Marta – where the average self-reported income was the lowest. The place that specified the highest average income from their farms was Santa Elena located in the core corridor.

As evident in Fig. 5.1, there is a strong positive linear relationship between income and the Ecological Footprint ($r = 0.436$). This correlation is stronger than for the positive linear correlation evident between income and the carbon Footprint ($r = 0.316$). This suggests that a higher income is increasing the Footprint in the study area. This finding is supported by other studies (e.g., consumption as income dependent, Fiala 2008; addressed by Levett 1998; Minx et al. 2009; and the

Table 5.1 Descriptive statistics of study sample demographics

Location	Gender (Females: Males)	Average Age (Years)	Years at Farm	Farm Size (hectares)	Annual Income (US$)
Santa Elena (SE = 50)	13:37	59.2 ± 13.9, *n* = 47	27.2 ± 16.5, *n* = 42	5.7 ± 13.4, *n* = 46	2364.4 ± 3001.6, *n* = 23
Quizarrá (QA = 30)	11:19	53.6 ± 15.0, *n* = 30	26.5 ± 17.5, *n* = 30	5.4 ± 8.5, *n* = 30	1435.4 ± 1522.4, *n* = 13
Montecarlo (MO = 10)	3:7	53.9 ± 12.9, *n* = 10	20.5 ± 17.7, *n* = 10	1.6 ± 1.3, *n* = 9	476.0 ± 388.4, *n* = 5
San Francisco (SF = 17)	9:8	54.5 ± 17.0, *n* = 17	27.2 ± 19.2, *n* = 16	9.3 ± 22.4, *n* = 17	1087.5 ± 1995.3, *n* = 8
Santa Marta (SM = 3)	2:1	52.7 ± 17.9, *n* = 3	37.3 ± 32.1, *n* = 3	1.4 ± 1.0, *n* = 3	730.0 ± 664.7, *n* = 2
San Ignacio (SI = 10)	5:5	46.6 ± 15.3, *n* = 10	23.3 ± 18.7, *n* = 10	2.7 ± 2.5, *n* = 9	1175.0 ± 1260.6, *n* = 4
Corridor (CR = 120)	*43:77*	*55.4 ± 14.9*, n = *117*	*26.3 ± 17.7*, n = *111*	*5.5 ± 12.9*, n = *114*	*1641.5 ± 2293.5*, n = *55*

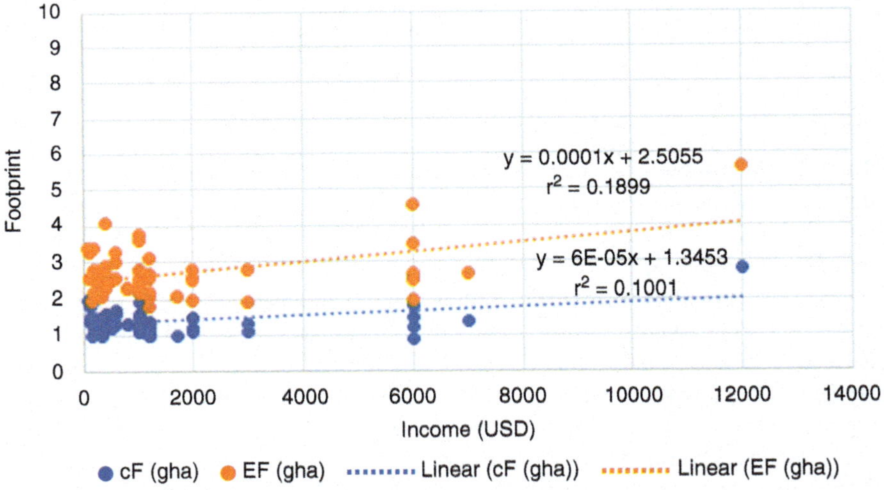

Fig. 5.1 Linear relationship between income (in US$) and the carbon (cF) and Ecological (EF) Footprints

assumption (consumption divided by income) made by Wackernagel 1998), that also found income to be associated with larger Footprints. Lenzen and Murray (2001) addressed this in some detail, suggesting that demographic factors (including income, expenditure, size, and the location of households) be examined since for an income increase of 10%, they found the Ecological Footprint to increase 3.8%. Moreover, they posited high-income consumption to be less 'land disturbing' than at low incomes (Lenzen and Murray 2001, p 244, Fig. 1) because the production of necessities (such as food and energy) is more land-disturbing than that of luxuries (such as recreation and financial services). In addition, household expenditure can be held as a better proxy for land use than income because not earning money and spending it exerts environmental pressure. Furthermore, the Ecological Footprint is lower in larger households due to the effects of sharing.

How does the Ecological Footprint account for reuse, recycling, and shared things? It only seems to entail the impact of reduced (or not) consumption. In Portugal, for example, Galli et al. (2020) found Almada to have the highest per person Footprint in 2016 likely due to its proximity to Lisbon, as the capital city and region in Portugal with the highest incomes, affecting the spending power of residents (Almada had the highest purchasing power attributable to tourism). Similarly, van den Bergh and Verbruggen (1999) considered changing income to have an indirect effect on consumption through changes in spending.

5.3 Farmers' Surveys

These surveys were based on the Footprint Calculator. For the six components based on land type of the Ecological Footprint (Fig. 5.2), the results convey the greatest standard deviation for grazing land and fishing grounds (Table 5.2), which reflects the disparate responses in these two components (see Table 5.2). The most critical components among the sampled population were the carbon Footprint and crop land (see Fig. 5.2) – which is perhaps unsurprising for farmers. However, the carbon Footprint was higher than expected, and it was sensitive to transportation

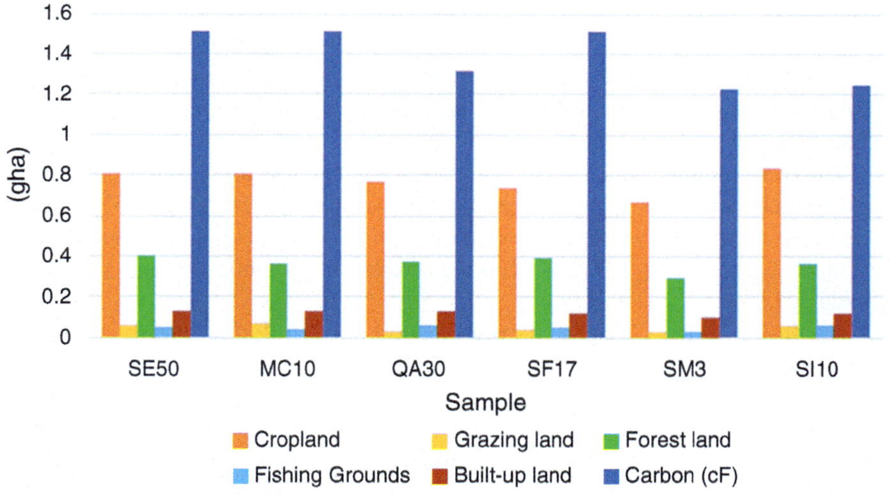

Fig. 5.2 Components of the Ecological Footprint for the towns in the study area

Table 5.2 Summary findings for the six Ecological Footprint components

Sample	Crop land	Grazing land	Forest land	Fishing grounds	Built-up land	Carbon (gha)
SE50	0.8 ± 0.4	0.1 ± 0.1	0.4 ± 0.1	0.1 ± 0.1	0.1 ± 0.1	1.5 ± 0.9
QA30	0.8 ± 0.4	0.0 ± 0.1	0.4 ± 0.1	0.1 ± 0.1	0.1 ± 0.0	1.3 ± 0.3
MO10	0.8 ± 0.4	0.1 ± 0.1	0.4 ± 0.1	0.0 ± 0.1	0.1 ± 0.1	1.5 ± 0.5
SF17	0.7 ± 0.2	0.0 ± 0.1	0.4 ± 0.1	0.1 ± 0.1	0.1 ± 0.0	1.5 ± 0.4
SM3	0.7 ± 0.1	0.0 ± 0.1	0.3 ± 0.1	0.0 ± 0.1	0.1 ± 0.0	1.2 ± 0.2
SI10	0.8 ± 0.5	0.1 ± 0.1	0.4 ± 0.1	0.1 ± 0.1	0.1 ± 0.0	1.3 ± 0.4
Average	*0.8 ± 0.3*	*0.1 ± 0.1*	*0.4 ± 0.1*	*0.1 ± 0.1*	*0.1 ± 0.1*	*1.5 ± 0.5*

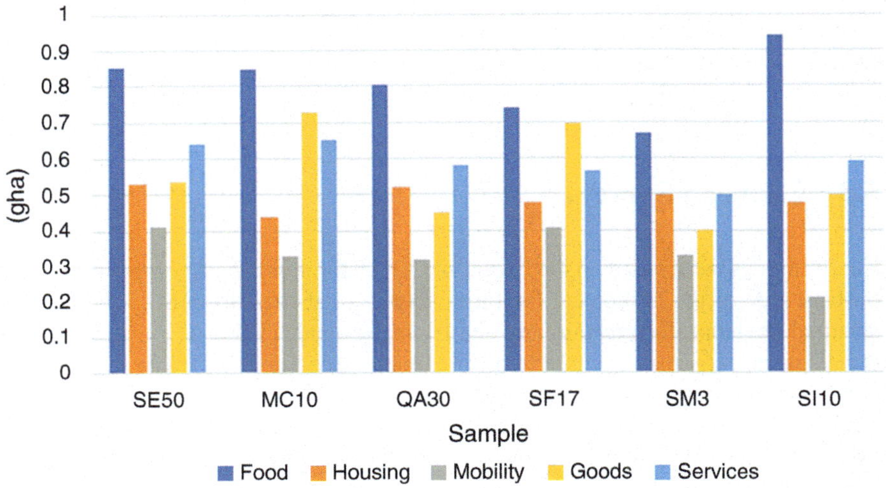

Fig. 5.3 Consumption categories for the towns in the study area

(e.g. distance travelled) – and the standard deviations were also greatest for these components, indicating more varied responses relating to them than the other components (see Table 5.2).

The consumption category with the highest average was food, followed by goods and services and housing, with transportation or mobility last (Fig. 5.3). Of these categories, responses were most varied concerning food, goods, and transportation or mobility (and least varied – or more consistent – for housing and services) – Table 5.3.

According to the Global Footprint Network (2020), Earth Overshoot Day for Costa Rica was August 10, 2020 (see Sect. 2.1). The result was similar to the corridor average (August 17, 2020) acquired for households in this sample. According to the Global Footprint Network (2020), Costa Rica's Ecological Footprint was 2.7 global hectares per person in 2016 (compared to 2.8 gha per person for the world then). By comparison, in Table 5.4, the towns in the study area within the corridor are overall very similar. The Ecological Footprint of participating households in the study area was 2.9 global hectares (Fig. 5.4), with around half of this

Table 5.3 Summary findings for consumption categories of each town

Sample	Food	Housing	Mobility	Goods	Services
SE50	0.9 ± 0.6	0.5 ± 0.2	0.4 ± 0.7	0.5 ± 0.4	0.6 ± 0.3
QA30	0.8 ± 0.6	0.5 ± 0.2	0.3 ± 0.2	0.5 ± 0.4	0.6 ± 0.1
MO10	0.9 ± 0.7	0.4 ± 0.2	0.3 ± 0.3	0.7 ± 0.7	0.7 ± 0.1
SF17	0.7 ± 0.4	0.5 ± 0.2	0.4 ± 0.3	0.7 ± 0.4	0.6 ± 0.1
SM3	0.7 ± 0.2	0.5 ± 0.3	0.3 ± 0.3	0.4 ± 0.5	0.5 ± 0.0
SI10	0.9 ± 0.8	0.5 ± 0.2	0.2 ± 0.2	0.5 ± 0.4	0.6 ± 0.2
Average	*0.8 ± 0.6*	*0.5 ± 0.2*	*0.4 ± 0.4*	*0.6 ± 0.5*	*0.6 ± 0.2*

Table 5.4 Summary of the Footprint Calculator results for each town

ID	Earth Overshoot Day	No. Earths	EF (gha)	cF (CO_2 emissions, tonnes/year)	% Total EF
SE50	20-Aug 2020 ± 69.1	1.8 ± 0.8	3.1 ± 1.4	4.4 ± 2.5	49.4 ± 6.3
QA30	03-Sep 2020 ± 62.9	1.6 ± 0.5	2.7 ± 0.8	3.8 ± 0.9	49.2 ± 5.7
MO10	08-Aug 2020 ± 55.7	1.8 ± 0.4	3.0 ± 0.7	4.4 ± 1.3	50.2 ± 9.0
SF17	08-Aug 2020 ± 37.4	1.7 ± 0.3	2.9 ± 0.5	4.4 ± 1.1	52.2 ± 6.8
SM3	16-Sep 2020 ± 39.6	1.3 ± 0.1	2.4 ± 0.4	3.5 ± 0.6	50.0 ± 3.5
SI10	06-Sep 2020 ± 65.6	1.6 ± 0.6	2.7 ± 1.1	3.7 ± 1.2	47.3 ± 6.1
Average	*17-Aug 2020 ± 56.3*	*1.7 ± 0.5*	*2.9 ± 0.8*	*4.2 ± 1.5*	*50.3 ± 7.0*

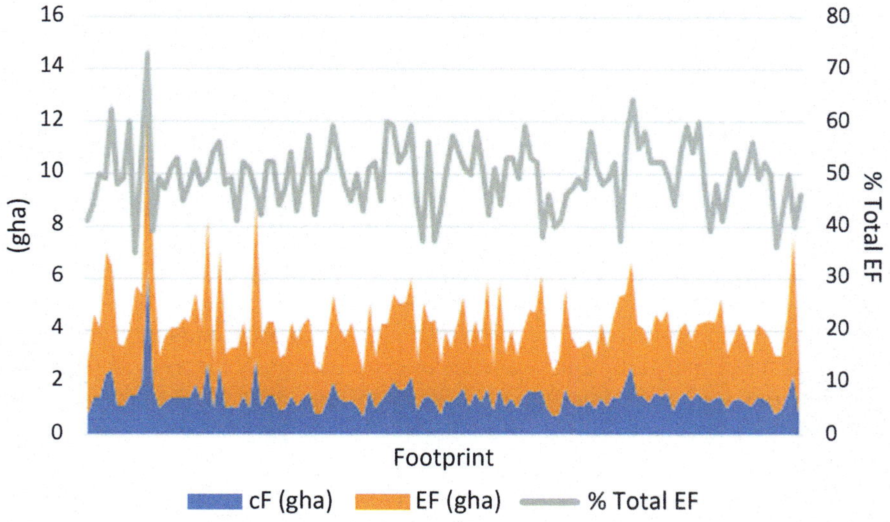

Fig. 5.4 Linear relationship between the carbon (cF) and Ecological (EF) Footprints

total attributable to carbon emissions (on average 4.2 tonnes/year of CO_2). The standard deviation around these values tended to be low, especially for the Ecological Footprint. This indicates that there was not very much disparity in the household Ecological Footprint between the corridor towns sampled, with averages between 2.4 and 3.1 global hectares at the extreme, respectively for Santa Marta and Santa Elena – a difference in the averages amounting to 0.7 global hectares. These results suggest that these small-scale farmers are not living within 1 Earth, as their average score was 1.7 Earths (see Table 5.4) – this is comparable to Costa Rica's 1.7 Earths for 2020 (Global Footprint Network 2020; see Table 3.1). This finding is surprising because one would expect small-scale farmers to have less of an environmental impact as producers, which helps to balance their consumption. However, since the Footprint Calculator does not input information for biocapacity (e.g., as evident for the carbon component), perhaps there is overestimating of their environmental impact and hindering their measured overall environmental performance.

5.4 Follow-up Interviews

The results of the follow-up interviews are summarised next – based on the English translation provided by the Spanish-speaking surveyors who helped to execute the fieldwork (see Acknowledgements). In this section, each of the seven (open-ended) interview questions will be portrayed individually in terms of their findings.

5.4.1 *Q1: What Type of Crops Do You Grow, and in What Proportion?*

There were four main or primary crops in the study area. They were coffee, followed by cattle (cows), and then grass (for cattle) as well as sugarcane (Fig. 5.5). Secondary crops included lychees, pepper, and cacao as well as fruits and vegetables and so on – in quantities of <2 hectares (except for SF02, where 3 hectares of land were being protected). These constituted a total of 267.2 hectares. Commonly, coffee was grown as the primary crop and sugarcane or cows with grass for cows as secondary crops. Crops were sometimes mixed, with up to four crops listed; but more often there were monocultures (in 75 cases, or 78%), where only one crop was grown – usually coffee (Fig. 5.6). It should be noted, however, that pastoral farming (mostly cows and some pigs) sometimes occurred alongside arable farming. It was less common to grow a mixture of crops beyond four crops, although this did occur in the study area.

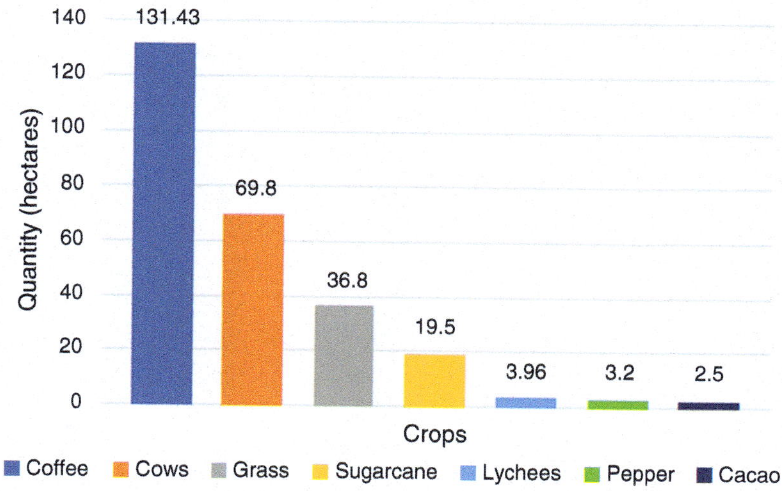

Fig. 5.5 Quantities of crops grown in the study area reported by farmers in the corridor ($\sum = 267.2$ hectares, $n = 107$)

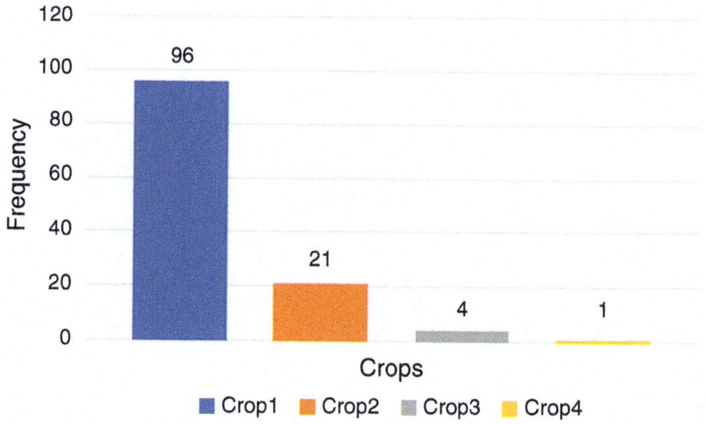

Fig. 5.6 Frequency counts of (any) crops grown in the study area ($n = 122$)

5.4.2 *Q2: How Much Do You Grow in a Year?*

It should be noted first that 2019 was used as the year for these estimates. Coffee was the easiest crop to monitor in terms of annual yields (Table 5.5). A total for the study area was 2927 *fanegas* ($n = 77$), where 1 *fanega* or bushel = 46 kilograms or about 100 pounds (a total of 2927 kg or 6452 lb). This was grown in a total of 142 hectares ($n = 76$), for a yield of 21 kilograms per hectare or 45 pounds per hectare. An average of 38 ± 54 *fanegas* (or 1748 kg = 3854 lb) of coffee were produced in 2019 based on self-reported data from the collected sample. They were grown on an

Table 5.5 Crop yields self-reported for 2019 in the study area

	Fanegas	Sugarcane (lb)	Income (US$)	Land size for coffee (hectares)	Total income (US$)
Total	2926.5	673,775.0	11,700.0	142.4	81,080.0
Average	38.0	61,252.3	1063.6	1.9	1725.1
StDev	53.5	104,187.6	1071.9	2.1	2464.9
n	77	11	11	76	47

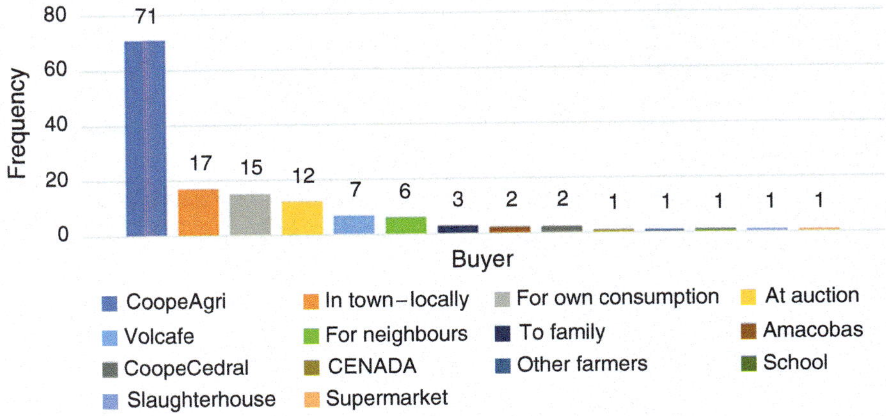

Fig. 5.7 Buyers of participating households' produce ($n = 140$)

average of 2 ± 2 hectares of land. This makes for an average yield intensity of 0.54 *fanega* per hectare, or roughly 25 kilograms or 55 pounds per hectare. The standard deviations, however, are greater than the averages – so, there is much disparity in the average yield intensity of coffee crops.

The income generated from this production totalled US$81,080 ($n = 47$). On an average household basis, this amounts to US$1725. This means that these farming households are living on <US$2000 per year based on their agricultural production for coffee. For this reason, other crops often accompany this primary crop, as demonstrated in Table 4.2, for example cows and/or grass. Some (newer) crops that are becoming more popular are lychees, peppers, and roots; however, more traditional crops such as cacao, bananas, plantains, citrus, and other fruits and vegetables are still planted. Chickens are sometimes also kept to produce eggs – either for household consumption or to be sold.

5.4.3 *Q3: Where Do You Sell Your Crops?*

Approximately half of participants (51%) mentioned that they sold their crops to CoopeAgri (Fig. 5.7), although there were a few that referred to other cooperatives in

Table 5.6 Summary of Fair Trade and organic farming results ($n = 120$)

Yes	No	Does not know
21 (17.5%)	96 (80.0%)	3 (2.5%)

the area. For example, Volcafe was referred to in Montecarlo (5×), Quizarrá (1×), and Santa Marta (1×). CoopeCedral (1× in San Francisco and 1× in Santa Marta) and others, such as CENADA, were also mentioned. Animals were sold at auction (9%). In some cases, products were sold locally to other people (12%) and to family and neighbours or other farmers. A woman's organisation (Amacobas) also supported agricultural production in at least a couple of cases in Quizarrá. In one known instance, produce (vegetables) was being sold at the supermarket and local school. Sometimes people were producing – either partly or wholly – for their own consumption (11%).

5.4.4 *Q4: Are You Participating in Fair Trade or Organic Farming?*

Eighty percent of the farmers who participated in this study responded 'No' to this question. Only 18% of farmers participated in some way (either partly or wholly). Most of them needed this question to be elaborated and the author suspects that more than the three participants noted in Table 5.6 were unfamiliar with 'Fair Trade' in particular. This could be because Fair Trade is only possible with cooperatives, so that a single family cannot usually be certified Fair Trade. Therefore, they were not engaging with Fair Trade as a household, or perhaps they did not recognise it by name (as Fair Trade).

Some farmers did express an interest in practising organic farming. However, as stated by SE01: 'Organic products need more money, time, and people and [they are] more difficult to sell, as there is no market to sell locally'. Another participant (QA13) wanted to establish an organic farm, so did not use a lot of chemicals, but found that it is difficult; and hoped that in 2 years they could produce organic vegetables. Others (e.g., QA17), were of the opinion that, 'there is no profit to produce organic coffee'. Another participant (QA19) noted that, 'it is better, but no; it is hard to be organic, [and] people need convincing'. For example, in the case of SF12 – who expressed that, 'if [one] wants to make land organic, [they need to produce] less because trees grow so fast; so, it is necessary to use both organics and chemicals'. At least six individuals desired to become more organic in their farming practice, while some 10 already used organic farming.

Table 5.7 Summary of fertiliser and pesticide use results ($n = 120$)

Yes	No	Does not know
84 (70.0%)	35 (29.2%)	1 (0.8%)

5.4.5 *Q5: Do You Use Fertilisers or Pesticides?*

Seventy percent of those interviewed reported using some proportion of chemicals to enhance their farming. Nearly a third (29%) used natural (nonchemical) approaches. Only one case seemed to lack an understanding of this question, so that evidently most farmers were aware of this issue (Table 5.7). In their responses to this question, participants expressed a desire to use organic fertilisers such as manure. They usually qualified a 'Yes' response by saying 'not so much' (SE14) and that they 'try to use as much as needed' and no more because they are 'trying to be organic'; in other words: 'A little fertilisers and pesticides when needed; trying to be organic' (SE40). Also, for example: 'Yes, [but] tries to change; wants to protect the environment' (SE20). Some reduce their frequency of use, if not the quantity, for example (SE26): 'Yes, not often – when needed (2 or 3 times per year)' and (SE50): 'Try using only fertilisers, but sometimes use pesticides (1 per 1.5 years)'. Others provided some explanation of reduced use: 'Yes, both – but minimum; [trying] to protect the environment, e.g. using trees and using less chemicals' (MO07). Some explained that, 'it is hard to farm without pesticides' (SE46).

The *Hemileia vastatrix* 'roya' fungus is another problem that they resolve through pesticide use: '[Try] to reduce chemicals, but bacteria comes twice per year' (SE48). Another person mentioned that: 'Yes – this year because of fungus problem; people know about the bad of pesticides; the community tries to Blue Flag and think about the next generation' (MO09). The Blue Flag programme was also referred to by another participant (QA16) and, therefore, merits explanation.

Costa Rica's 'Blue Flag Ecological Program' is concerned with water quality, and was originally established in 2002 to protect the country's beaches from pollution (source: https://web.archive.org/web/20160412220826/http://www.visitcostarica.com/ict/paginas/mapas/areasurf.asp). It allows for multi-stakeholder engagement, including citizen participation. Among its assessment criteria are water-related quality (affecting oceans, coasts, and potable water) as well as environmental education plus security and administration. This means that these participants are concerned about pollution-related issues that affect their water supply, including garbage, industrial waste, and runoff water (as into rivers).

One participant elaborated that: 'No – [does] not use chemicals for almost 15 years, but [uses] natural things to keep clear of bugs and fungus; keep trees by themselves' (QA08). They were more inclined towards using fertilisers than pesticides, as for example (QA03): 'Yes, uses fertilisers, but less so pesticides'; similarly (SF10): 'No fertilisers, but ... a little pesticides (rarely)'. Another farmer noted: 'Rarely – leaves the grass that cuts behind as ground cover and to fertilise crops' (SF08). Organic fertilisers were used by at least four participants, with one mentioning the use of 'manure for fertiliser' (SI07).

5.4.6 *Q6: Do You Benefit from Payments for Ecosystem Services?*

By far, the most incomprehensible question was this one (followed by Fair Trade in Q4). The interviewer had to explain – with few exceptions – what this meant in their own words at almost all of the follow-ups. In reality, this question should have been omitted from the study, that concerned only small-scale farmers. Payments for Ecosystem Services is practiced more by large landowners, since payments are based on the acreage forested that is being set aside – so that land size matters. Most people (106 participants or 88%) answered 'No' here without further elaboration. Those that did say more, in some cases, provided commentary on unrelated aspects. This observation also indicates confusion in understanding this question.

Those who responded positively tended to connect it to CoopeAgri. One person explained that: 'Some people came long time ago – they gave some trees to replant and 100,000 colones in equipment' (SE40) – also (SI08): 'A few years ago, they gave some trees (likes to have trees)'; similarly (MO10): 'CoopeAgri gave . . . some trees'. One individual mentioned that he: 'Participated with CoopeAgri for conservation of [the] Penas Blancas River' (QA16); likewise: 'A few years – with CoopeAgri' (SF06). A participant was closely aware of it: 'No, but . . . knows about that because [works] for the cooperative' (SI06). Another participant commented that: 'No, but Amacobas produce 500 trees per year for Blue Flag' (QA09). In a couple of instances, those participants who admitted to not knowing about this, wished to learn more (e.g., SI10).

5.4.7 *Q7: Have You Ever Been Approached Before to Complete a Survey or an Interview?*

Some 26 participants (22%) had not been interviewed before and this was their first interview. It was important to ascertain this because of the phenomenon of 'interview exhaustion' that could occur within a community such as this one that is linked to a university. A few people noted that Los Cusingos had dropped by as well as more commonly students – mostly from York University.

5.5 Conclusions

This chapter has provided some reassurance that the case study brings a novel contribution by reaching almost a quarter of new participants in the study area. It also conveys a lack of familiarity among case study participants regarding concepts such Payments for Ecosystem Services and Fair Trade. Perhaps this is the product of individual familiarity of farmers, however, and does not really reflect the knowledge of policymakers and (educated) town officials. The local context is extremely important here, as most participants in the case study were uneducated and reliant on outreach from experts from places such as York University, Los Cusingos, CoopeAgri, and so on.

Although some of the main findings have been presented in this chapter, the next chapter includes other materials derived from the survey-interviews that have not been addressed so far in this brief. Furthermore, a discussion of the implications of the findings and consideration of the potential solutions will ensue (Chap. 6). As revealed in the current chapter, though, the Ecological Footprint of study area households comprising small-scale peasant (*campesino*) farmers was on average greater than the national value published in the 2020 Edition of the National Footprint Accounts. This implies that aspects of their lifestyles need to be scrutinised and suggestions made for (behavioural) change, assuming that the results are revealing of their lifestyles and not attributable to error (which will be considered next).

References

Fiala N (2008) Measuring sustainability: why the ecological footprint is bad economics and bad environmental science. Ecol Econ 67(4):519–525. https://doi.org/10.1016/j.ecolecon.2008.07.023

Galli A, Iha K, Moreno Pires S, Mancini MS, Alves A, Zokai G, Lin D, Murthy A, Wackernagel M (2020) Assessing the ecological footprint and biocapacity of Portuguese cities: critical results for environmental awareness and local management. Cities 96:102442. https://doi.org/10.1016/j.cities.2019.102442

Global Footprint Network (2020) National Footprint Accounts. https://data.footprintnetwork.org/#/

Lenzen M, Murray SA (2001) A modified ecological footprint method and its application to Australia. Ecol Econ 37(2):229–255. https://doi.org/10.1016/S0921-8009(00)00275-5

Levett R (1998) Footprinting: a great step forward, but tread carefully — a response to Mathis Wackernagel. Local Environ 3(1):67–74. https://doi.org/10.1080/13549839808725545

Minx JC, Wiedmann T, Wood R, Peters GP, Lenzen M, Owen A, Scott K, Barrett J, Hubacek K, Baiocchi G, Paul A, Dawkins E, Briggs J, Guan D, Suh S, Ackerman F (2009) Input-output analysis and carbon footprinting: an overview of applications. Econ Syst Res 21(3):187–216. https://doi.org/10.1080/09535310903541298

Van den Bergh JCJM, Verbruggen H (1999) Spatial sustainability, trade and indicators: an evaluation of the 'ecological footprint. Ecol Econ 29(1):61–72. https://doi.org/10.1016/S0921-8009(99)00032-4

Wackernagel M (1998) The ecological footprint of Santiago de Chile. Local Environ 3(1):7–25. https://doi.org/10.1080/13549839808725541

Chapter 6
Implications

Abstract It is important to consider the ramifications of the fieldwork study findings for the sampled study area in comparison to Costa Rica at the national level and globally to the world. This chapter revisits the fieldwork results from the study area in the Alexander Skutch Biological Corridor and examines the solutions presented by the Footprint Calculator for the categories of city, energy, food, and population. It is important to do this because local–regional lifestyle or cultural trends in the corridor operate to lower the Ecological Footprint (or augment biocapacity), offering lessons learned or solutions. The chapter then draws upon the lessons learned from these solutions as well as uniquely from the study area based on the fieldwork findings. As part of this assessment is a sensitivity analysis based on three scenarios stemming from questions in the Footprint Calculator. The findings reveal housing to be the most affected in the sensitivity analysis, followed by transportation, and lastly by food categories. Finally, the contributions of the field study as a case study justify the research and present the implications of the case study.

Keywords Case study · Field study · Data quality · Lessons learned · Sensitivity analysis · Cultural implications · Lifestyle choices · Rural environment · Socioeconomics

It is important to consider the ramifications of the fieldwork findings for the sampled study area in comparison to Costa Rica at the national level and globally to the world. This chapter revisits the fieldwork results from the study area in the Alexander Skutch Biological Corridor and examines the solutions presented by the Footprint Calculator for the categories of city, energy, food, and population. It is important to do this because some local–regional lifestyle or cultural trends in the corridor may operate to lower or augment the Ecological Footprint (and biocapacity), offering lessons learned or solutions.

When compared to the world's Ecological Footprint of 2.8 global hectares per person for 2016 (Global Footprint Network 2020), Costa Rica's Ecological Footprint is similar at 2.7 global hectares per person. To achieve strong sustainability,

M. J. Thornbush, *The Ecological Footprint as a Sustainability Metric*,
SpringerBriefs in Environmental Science,
https://doi.org/10.1007/978-3-030-62666-2_6

nations need to maintain a low Ecological Footprint and maximise their biocapacity in order to avoid getting into deficits and ecological overshoot. Overall, Costa Rica has a biocapacity that is sufficiently large (1.6 gha per person, the same as for the world's biocapacity per person, according to the Global Footprint Network 2020) to support its current Ecological Footprint, so that it is in a slight deficit (of −1.1 gha/person). This deficit is likely greater than the anticipated deficit for the study area in the Alexander Skutch Biological Corridor. The reason for this is that these are producing households that grow crops and, thereby, their (arable farming) production activities should curtail their carbon Footprint, which is a substantial part of their total Ecological Footprint. Nevertheless, more cows (pastoral farming) in the corridor could further contribute to methane release – although this waste product is not specifically measured by the Ecological Footprint and biocapacity accounting at present.

According to Lin et al. (2018, p 9), Earth's ecological overshoot commenced in the 1970s and continues to grow at an average rate of 0–4% per year. For Costa Rica, as examined in the previous chapter (Chap. 5), ecological overshoot began in 1991, some 20 years later. Ecological overshoot is indicative of the depletion and degradation of natural capital accompanied by an accumulation of wastes (in the atmosphere and elsewhere). Ecological overshoot can be seen through deforestation, biodiversity loss, soil degradation, fisheries depletion, and climate change (Lin et al. 2018).

This chapter revisits the findings from the previous chapter (Chap. 5) in an overview and discussion of the current state of the study area in the Alexander Skutch Biological Corridor compared to the national situation for Costa Rica. Solutions are presented and considered in view of the fieldwork results. The lessons learned from these solutions as well as stemming from the fieldwork are outlined and scrutinised, including a sensitivity analysis. Finally, the fieldwork contributions appear last in this chapter before the conclusions.

6.1 Results for the Corridor

From the previous chapter (Chap. 5), the following points can be gleaned:

- *Demographics* – mostly men in their mid-50s participated in the survey-interviews in the corridor case study. The average participant resided on their farm for over 25 years and were landowners. Their land was on average approximately 5 hectares in size, and they reported making <US$2000 per year.
- *Farmers' surveys* – Out of 120 participants in this study, the average carbon Footprint was calculated at the household level to be <1.5 global hectares. The Ecological Footprint of participating households was on average almost 3.0 global hectares. These small-scale farmers scored highest on crop land as a land type and consistently lowest for the components grazing land and fishing grounds. The highest values were received on average for the consumption

category of food. On average, their Earth Overshoot Day (in the study area: August 17, 2020) roughly corresponded to the national result of August 10, 2020 (refer to Sect. 2.1). Households in the corridor required >1.5 Earths to meet their resource requirements.

- *Follow-up interviews* – Based on the seven supplementary questions asked, it was possible to discern the following:

  - *Q1: What type of crops do you grow, and in what proportion?* Coffee was by far the most frequently grown crop by these farmers, including mainly as a monoculture crop, but also as part of a mixed crop system that tended to include up to four primary 'staple' crops. Among their main produce was coffee as well cows and grass, plus sugarcane; followed by lychees, pepper, and cacao as secondary (and newly introduced) crops.
  - *Q2: How much do you grow in a year?* Farmers produced nearly 3000 *fanegas* (or bushels) in 2019 on almost 150 hectares of land. This resulted in approximately 138,000 kilograms or about 14 million pounds, an average annual yield of almost 1000 kilograms or 10,000 pounds per hectare. Their average stated annual income from this production was in the area of US$81,000, or < US$2000 per household, each year.
  - *Q3: Where do you sell your crops*? The top buyer of their produce – purchasing about half of their products – was by far CoopeAgri. However, there were another 13 buyers.
  - *Q4: Are you participating in Fair Trade or organic farming*? Although most farmers did not understand what 'Fair Trade' represented, they were enthusiastic about fair dealings with others as well as participating in organic farming.
  - *Q5: Do you use fertilisers or pesticides*? They preferred using fertilisers over pesticides, especially organic fertilisers (manure). The farmers understood the benefits of using both of these, but also considered harsh chemicals that are harmful to the environment. They commented that they worked with the Blue Flag Ecological Program wherever possible. Their use of artificial pesticides was reduced as much as possible, especially to combat the roya fungus; they were also cognisant of the need to spray as infrequently as possible in order to promote soil health.
  - *Q6: Do you benefit from Payments for Ecosystem Services?* This was the least developed of the responses in the follow-up interviews. Participants often gave simple 'Yes' or 'No' answers. Those that elaborated on their response sometimes veered off topic or referred to forestation efforts by Los Cusingos or CoopeAgri. They were largely unaware of any national benefits programme.
  - *Q7: Have you ever been approached before to complete a survey or an interview?* This final question in the follow-up interviews gauged the level of 'exhaustion' among the sampled participants. Almost a quarter of them were newly surveyed and had not been previously interviewed. Among those that had been interviewed before, they mentioned (York University) students and people from Los Cusingos and CoopeAgri.

6.1.1 Error Analysis

The percentage difference of reported income has been based on the demographics section enquiring about annual income versus information on yields provided in the follow-up interviews. The proportions have been calculated to assess the human error in reporting these figures – to ascertain the consistency of information. Based on a subsample of 10 samples (Table 6.1), where there was overlap in the responses, it was possible to determine an error derived using the following equation:

$$\%\text{difference} = \frac{\text{Absolute value of the change}}{\text{Average of the two numbers}} \times 100\% \qquad (6.1)$$

Half of the subsamples reveal no difference, indicating consistent reporting. However, the remainder of the values in the subsample add up to a difference of 405% because participants provided inflated values for the first value (V1) in the demographics portion of the questions – at the start of the survey-interviews. This could be because the second value (V2) was for specific yields and they did not provide all yields (for all crops) or, otherwise, their annual income is based on something other than the specific agricultural yields that they reported.

As for the standard error, it was possible to determine this based on the following equation:

$$\textit{Standard error} = \frac{\text{Standard deviation}}{\sqrt{\text{Sample size}}} \qquad (6.2)$$

Based on the previous chapter (see Chap. 5), for demographic information it is possible to discern that annual income had the highest standard deviation (StDev) values. When the standard error is calculated, it appears that the standard error is inflated for annual income – likely because of high initial reported income during the

Table 6.1 Calculations of % difference for reported income

ID	V1 (US$)	V2 (US$)	V1 - V2	(V1 + V2)/2	[V1 - V2/(V1 + V2)/2] × 100%
SE08	200	200	0	200	0.0
SE16	400	400	0	400	0.0
SE17	3000	1200	1800	2100	85.7
SE18	400	400	0	400	0.0
SE20	3000	3000	0	3000	0.0
MO02	80	80	0	80	0.0
MO04	220	100	120	160	75.0
MO05	1800	1000	800	1400	57.1
SF01	800	200	600	500	120.0
SF02	800	400	400	600	66.7
Total	*10,700*	*6980*	*3720*	*8840*	*404.5*

Table 6.2 Standard error of self-reported annual income

Location	Annual Income (US$)	Standard Error
Santa Elena (SE = 50)	2364.4 ± 3001.6, *n* = 23	625.9
Quizarrá (QA = 30)	1435.4 ± 1522.4, *n* = 13	422.2
Montecarlo (MO = 10)	476.0 ± 388.4, *n* = 5	173.7
San Francisco (SF = 17)	1087.5 ± 1995.3, *n* = 8	705.5
Santa Marta (SM = 3)	730.0 ± 664.7, *n* = 2	470.0
San Ignacio (SI = 10)	1175.0 ± 1260.6, *n* = 4	630.3
Corridor (CR = 120)	*1641.5 ± 2293.5,* n = *55*	*309.3*

Table 6.3 Standard error of the Ecological Footprint results

Location (n)	Earth Overshoot Day (StDev)	Standard Error	% Total EF (StDev)	Standard Error
SE50	69.1	9.8	6.3	0.9
QA30	62.9	11.5	5.7	1.0
MO10	55.7	17.6	9.0	2.9
SF17	37.4	9.1	6.8	1.7
SM3	39.6	22.9	3.5	2.0
SI10	65.6	20.8	6.1	1.9
CR120	*56.3*	*5.1*	*7.0*	*0.6*

demographics portion of the survey-interviews. It should be noted here that although the sample size was low in Montecarlo ($n = 10$) for the reported annual income, its standard error is still lowest among the towns and corridor standard error – conveying a consistency in responses.

For the Ecological Footprint, results were selected for a standard error check (Table 6.2) where standard deviation values were highest – based on Table 5.5 in the previous chapter. This error analysis conveyed that Earth Overshoot Day and % Total EF (of the cF) had the most inflated standard deviation values across all Ecological Footprint components and consumption categories. The average standard error for Earth Overshoot Day was the greatest of all at 5 days for all corridor towns sampled. The other calculated standard error result for % Total EF was <1%, as shown in Table 6.3.

Evidently, sample size did affect the standard error. It was not possible, however, to sample at least 50 people in each town because of temporal and monetary constraints on the fieldwork. Also, towns were not of equal sizes (Table 6.4, based on estimates from our field guide, Alex Fonseca). Based on these estimations, Montecarlo, San Francisco, and especially Santa Marta were underrepresented in this sample. The rationale for including the latter towns was to compensate for the small sample size in Montecarlo. The original plan was to sample only three towns located in or towards the core corridor, but this had to be expanded because of Montecarlo. Conversely, Santa Elena and Quizarrá were oversampled, and only San Ignacio had an adequate proportion of sampling executed. However, the error associated with over-sampling and under-sampling was balanced for the corridor.

Table 6.4 Estimated households in the sampled towns

Town sample	Estimated population size	Proportional sample size	Difference
SE50	450 (TD5 = 100) = 550	29 (6) = 35	50–35 = +15
QA30	250	16	30–16 = +14
MO10	200	13	10–13 = −3
SF17	350	23	17–23 = −6
SM3	350	23	3–23 = −20
SI10	150	10	10–10 = 0
Total	*1850*	*120*	*+29 - (−29) = 58*

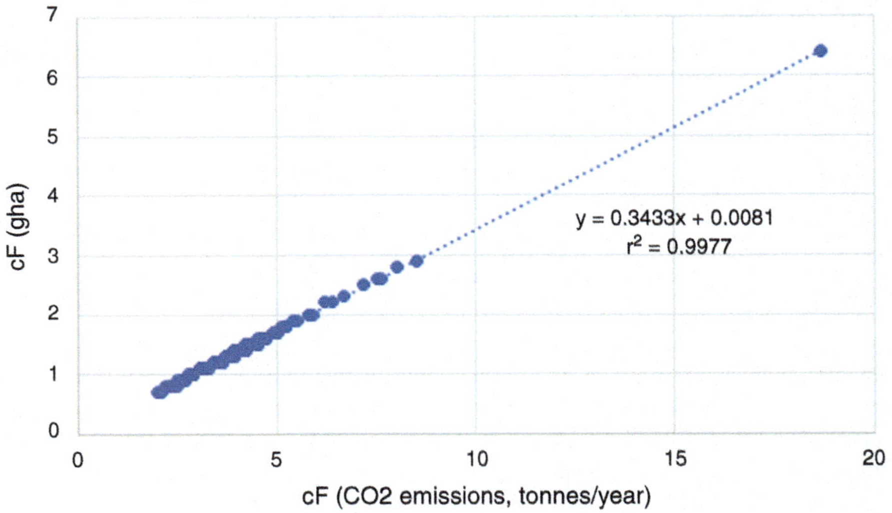

Fig. 6.1 Linear relationship between measurements of the carbon Footprint

Besides affecting the standard error, the small sample size, particularly in Montecarlo, San Francisco, and especially Santa Marta caused problems for performing parametric statistical testing (e.g., ANOVA) at town level, which would require at least 30 samples.

There is nearly a perfect linear relationship ($r = 0.999$) between the carbon Footprint outputs in CO_2 emissions (tonnes/year) and as a land-based measure (gha) – Fig. 6.1. This attests to the low error associated with the conversion between the two units. It was interesting to examine this because some authors (cf. Galli et al. 2012; Wiedmann and Minx 2008; Wright et al. 2011) have criticised or cautioned against the error associated with the land conversion (to gha) in the Ecological Footprint. However, in the current study, the linear relationship is almost perfect, indicating insignificant error attributable to this conversion from the CF (based on tonnes/year) to cF (expressed in gha).

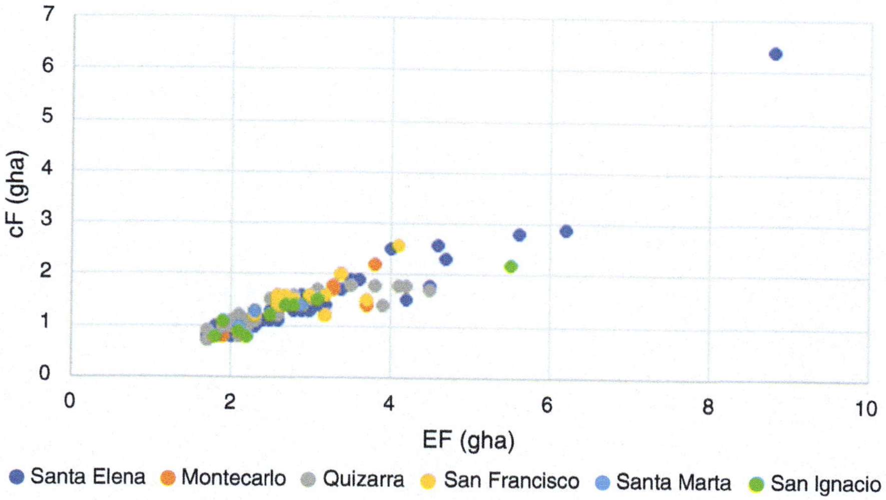

Fig. 6.2 Linear relationship between the Ecological Footprint and carbon Footprint

Additionally, a positive linear correlation ($r = 0.908$) was evident between the Ecological Footprint and its carbon Footprint component (Fig. 6.2). However, there is one outlier (SE12) that slightly augments the strength of this relationship (from $r = 0.888$).

6.2 Solutions

As part of the outputs from the Footprint Calculator, in addition to examining the results, there are solutions that need to be considered. These appear related to city, energy, food, and population themes (Box 6.1) that stem from the results (as an output), so will be addressed in this chapter. Of these, the most relevant to the research in the corridor is food. These are rural communities that still function based on unpaved country roads – although this is changing, as the main road through the corridor was being widened in preparation for paving at the time of the fieldwork (Fig. 6.3). This indicates that the corridor is urbanising, as evident through its connections with the road network, connecting it to other places and encouraging commuting with nearby cities, such as San Isidro del General. Additionally, farms are being sold and subdivided into housing lots.

Fig. 6.3 Paving of the main road through the corridor. (Alexander Skutch Biological Corridor, photo by M. Thornbush, March 2020)

Box 6.1 Solutions

City – With 70–80% of the world's population expected to live in cities by 2050, smart urban planning and development strategies are crucial to managing our resources.

Visit your city's website and challenge your city leaders to support sustainability policies.

More information: http://www.overshootday.org/take-action/cities/

Energy – Renewable energy is a direct path to reducing your Ecological Footprint and addressing climate change.

Can you take transit, bicycle or walk instead of driving solo at least once a month? Once a week?

More information: http://www.overshootday.org/take-action/energy/

***Food** – Diet and cutting food waste are powerful sustainability levers.

More information: http://www.overshootday.org/take-action/food/

Can you be a smarter shopper and reduce food waste? Can you try a new vegetarian recipe once a month? Once a week?

(continued)

Box 6.1 (continued)

Population – Addressing population size is essential to creating a sustainable future for all within our planet's ecological budget.

You can choose the size of your family to affect our long-term Footprint. Support women's rights and access to family planning.

More information: http://www.overshootday.org/take-action/population/

Transcribed by Mary Thornbush, Faculty of Environmental Studies, York University (Email: mthornbu@yorku.ca) based on: http://www.footprintcalculator.org/.

Some 95% of energy is renewable in Costa Rica (e.g., Delahaye 2019). This should make for a reduced Footprint; however, transportation is still a consideration here, as most people use inefficient private vehicles to get around town. Additionally, most people in this study did not rely on public transportation (buses, as there is no train) and a majority did not bicycle – although some were seen to walk locally. Almost none of the study participants travelled regularly each year by airplane. Most of their driving was done on a 'solo' basis because of a lack of a car-share scheme in the study area. The household carbon Footprint (cF = 1.5 gha) represents about half of the Ecological Footprint. This is mostly attributable to driving private vehicles, especially where farm locations are separate (sometimes distant) from the place of residence. Rather than living in town, perhaps it is more prudent for some of these families to live on a farm homestead and use public transportation (bus) more often, where possible. One 50-year-old woman (QA30), for example, stated that she had to walk 2 kilometres to get to the nearest bus stop in Quizarrá.

As regards population, there are >500 farming households in the corridor and > 2000 people (see Table 6.4). Population growth is steady, with most couples having up to two children. There are women's organisations in the area, including Amacobas that supports women's rural livelihoods.

The area that really needs focus to reduce the household Ecological Footprint in the study area (from an average 1.7 Earths) is wrapped up in food. There is a lack of variety in the meals that some of these *campesinos* are having because they depend on what is grown locally or even on their own farms. It should be noted, however, that there are local stores in these towns where food items can be purchased. Although these foods tend to be locally sourced, not all of them are organic and most receive chemical sprays from the farmers. Traditional diets have an abundance of beans, rice, and corn as well as eggs. Meat is normally ingested at least once per week and a minority of farmers eat fish and seafood. However, generally, the grazing land component is low in the participating households, so perhaps meat consumption is not at issue here. Food scraps are composted by households. There is a free recycling programme, but household waste costs for disposal. Some farmers, however, are either burying or burning non-compostable waste, and this behaviour could be improved with educational outreach activities.

6.3 Lessons Learned

Households that were living 'in harmony' with the environment had Ecological Footprints of 1 Earth (or 1.7 gha), but there were no cases observed to have less than this value. These cases, where participants were living at 1 Earth, can be construed as exemplary, therefore, where lessons can be learned based on the fieldwork results. The data include information on Earth Overshoot Day for 2020 (EOD 20) as well as components of the Ecological Footprint (crop land = CL, grazing land = GL, forest land = FL, fishing grounds = FG, build-up land = BL, and carbon = cF) and consumption categories (food = F, shelter or housing = H, mobility = M, goods = G, and services = S) in addition to the Ecological Footprint (EF, gha) and carbon Footprint (CF, CO_2 emissions, tonnes/year), including its proportion (%) of the Ecological Footprint (Table 6.5a). Three cases in Santa Elena (namely, SE39, SE40, and SE47) are noteworthy, and Quizarrá had two (namely, QA09 and QA19). These five exemplary cases will be taken in turn and analysed so that lessons can be learned based on this information.

These cases (at 1 Earth) had Earth Overshoot Days later in the year, in December or January 2021 of the following year, meaning that resources are available throughout the entire year and even into the next one. Crop land is within 0.6 global hectares and forest land within 0.4. Both grazing land and built-up land are consistently zero and 0.1, respectively; and fishing grounds are also low at 0.1 global hectares. The carbon Footprint based on global hectares (cF) is kept within 0.9 (<1 gha), with CO_2 emissions kept within 2.6 tonnes per year (CF). The proportion of the Ecological Footprint attributable to the carbon Footprint (cF) is within 51%, so constitutes approximately half of the Ecological Footprint. Consumption categories pertaining to food, housing, mobility, goods, and services are respectively within 0.8, 0,7, 0.4, 0.1, and 0.5 global hectares. The household Ecological Footprint itself is within 1.8 global hectares. This suggests that a household Ecological Footprint of 1.8 global hectares can still be considered as living within 1 Earth. However, because in 2016 there were 1.6 global hectares of biocapacity on the planet (see Table 3.1, based on the National Footprint Accounts, 2020 Edition; Global Footprint Network 2020), our Ecological Footprint should be constrained to within that amount of biocapacity – to sustainably support our activities.

By comparison, six cases with a worse Ecological Footprint of 2 Earths appear in Table 6.5b. These cases represent an Ecological Footprint that is exactly one magnitude higher than those in the previous table (see Table 6.5b). It is evident for these cases that the carbon and Ecological Footprints have the most inflated results (the greatest difference), and for the consumption categories it is goods (see Table 6.5).

Doubling the Ecological Footprint to 2 Earths increases the impact evident with 1 Earth, which indicates small-scale farmers in the corridor who have a low Ecological Footprint but are living in poverty. This means that having a low Ecological Footprint is environmentally beneficial, but has consequences for the standard of living (refer to discussion on the quality of life in, e.g., Holmberg et al.

Table 6.5 Cases with exactly (a) 1 Earth and (b) 2 Earths

(a) **1 Earth**

ID	EOD 2020	CL	GL	FL	FG	BL	cF	F	H	M	G	S	EF	CF	%
SE39	13 Dec	0.4	0	0.4	0	0.1	0.8	0.4	0.7	0.1	0	0.5	1.8	2.2	42
SE40	09 Jan[a]	0.5	0	0.2	0	0.1	0.8	0.5	0.4	0.3	0.1	0.4	1.7	2.4	50
SE47	27 Dec	0.6	0	0.2	0.1	0.1	0.7	0.8	0.3	0.2	0	0.4	1.7	2.1	43
QA09	22 Dec	0.5	0	0.2	0	0.1	0.9	0.5	0.4	0.4	0	0.4	1.7	2.6	51
QA19	08 Jan[a]	0.6	0	0.2	0.1	0.1	0.7	0.7	0.4	0.1	0	0.4	1.7	2	40

(b) **2 Earths**

ID	EOD 2020	CL	GL	FL	FG	BL	cF	F	H	M	G	S	EF	CF	%
SE10	26 Jun	0.8	0.1	0.5	0.1	0.2	1.9	0.8	0.6	0.5	0.9	0.7	3.5	5.4	53
SE42	02 Jul	0.5	0	0.6	0	0.2	2	0.2	1	0.9	0.6	0.7	3.4	5.8	59
SE48	01 Jul	0.8	0.1	0.5	0.1	0.2	1.7	0.9	0.6	0.4	0.8	0.6	3.4	5	51
MO02	30 Jun	0.6	0.1	0.5	0	0.2	2	0.3	0.5	0.2	1.7	0.7	3.4	5.9	59
MO04	04 Jul	0.8	0.1	0.4	0	0.1	1.8	0.7	0.3	0.1	1.6	0.7	3.3	5.2	54
SF01	28 Jun	0.6	0	0.6	0	0.2	2	0.4	0.7	0.5	1.1	0.7	3.4	5.8	58

[a] 2021

1999; Kitzes et al. 2009; Levett 1998; Wackernagel 1998). For this reason, it is necessary to consider the Ecological Footprint and biocapacity from an integrated sustainability approach that considers the environmental as well as socioeconomic implications. Another issue arising from living at 1 Earth is that health may be impacted, improved due to a cleaner environment (where there is a low environmental impact stemming from consumption); however, this effect could be counteracted if there are insufficient funds available to farmers for healthcare.

At exactly 2 Earths, the Earth Overshoot Day moves up in the year to June or July 2020 (so, at mid-year), meaning that resources are being depleted sooner in the year. Crop land increases up to 0.8 (from 0.6 gha), grazing land to 0.1 (from zero), forest land to 0.6 (from 0.4), fishing grounds 0.1 (same upper boundary), and built-up land to 0.2 (from 0.1) global hectares – see Table 6.5. The carbon Footprint component is now more than double (up to 2.0 gha), and its emissions are within 5.9 tonnes per year, so have doubled. The proportion of the Ecological Footprint represented by the carbon Footprint is up to 59%, an increase of 8% at the upper end – which is the least increase evident in the shift from 1 to 2 Earths, apart from fishing grounds. This suggests that the proportion of the carbon Footprint in the total Ecological Footprint and fishing grounds are the least responsive to doubling the number of Earths. The household Ecological Footprint itself increases up to 3.5 global hectares, nearly double.

6.3.1 Sensitivity Analysis

An exercise entailing 'sensitivity analysis' in this context is where variable changes are adopted from the suggested solutions in order to improve the household Ecological Footprint. The variables that could be altered here are as follows:

1. Question 1: Changing to a vegan diet (food);
2. Question 8: Reducing wastes and recycling (housing); and
3. Question 10: Increasing the use of public transport (transportation).

These changes are applied to a subsample of disparate households in the corridor (the first households in each town examined in the 1 Earth and 2 Earths analysis – see Table 6.5). The number of Earths resulting from these improvement scenarios appear in Table 6.6, and the other results are in Table 6.7.

It is evident that these changes allow the 1 Earth cases to be reduced to <1 Earth when a vegan scenario (1) is employed. It is noteworthy that a vegetarian or vegan diet may not be considered to be healthiest, but there is an effect on the Ecological Footprint observed in this study based on case results. Importantly, in all cases, it is possible to reduce the impacts to <2 Earths with the other scenarios, especially by reducing wastes and recycling (e.g., MO02). The tallied differences in Table 6.6 suggest that the changes employed in the housing scenario (2) are most impactful on the household Ecological Footprint – based on the number of Earths, and the least effective is the food scenario (1), with the transportation scenario (3) as middling in

Table 6.6 Scenarios and their effects on the number of Earths

Scenario	Variable Change(s)	ID	No. Earths	No. Earths	Difference (No. Earths)
1	Question 1: Changing to a vegan diet – No meat, eggs, and dairy; 'never' selected for all items	SE39	1	0.9	0.1
		QA09	1	0.8	0.2
		SE10	2	1.6	0.4
		MO02	2	1.8	0.2
		SF01	2	1.9	0.1
	1. Total difference = 1				
2	Question 8: Reducing wastes and recycling – No wastes generated, but recycle paper and plastics	SE10	2	1.1	0.9
		MO02	2	0.9	1.1
		SF01	2	1.2	0.8
	2. Total difference = 2.8				
3	Question 10: Increasing the use of public transport – No private vehicle used (car or moped), all km transferred to bus (as there is no train service available)	QA09	1	1	0
		SE10	2	1.8	0.2
		MO02	2	1.3	0.7
		SF01	2	1.7	0.3
	3. Total difference = 1.2				

its effectiveness. Put together, however, these scenarios can have substantial impacts (with a total difference of 5 Earths for the three improvement scenarios). Such an analysis could be extended to include more improvement scenarios that entail more questions in the different parts of the Footprint Calculator.

6.4 Contributions

The fieldwork component contributed towards supporting youths in the corridor during paid surveying in February to March 2020. The fieldwork findings reveal the need for further workers to engage with this community in outreach activities to educate them about food alternatives (e.g., vegetarian or vegan diets) to reduce their food Footprint and the crop land component. Their use of pesticides is still problematic, as evidenced by the lack of protective garments and gear worn by farmers spraying chemicals. More research needs to be devoted to exploring the impact that this has on cancer rates and other health implications.

In addition, some of the added comments during the survey-interviews revealed a need to educate communities in the corridor about recycling and improved waste management. Although, some people already reuse items, such as tires, plastics bottles, metal, and more (Fig. 6.4), some people are still burning their garbage (e.g., paper) and/or burying it (putting it in a hole) based on the perception that, 'if you cannot see it, then it's okay' (QA20). This has implications for their carbon Footprint, as does too the lack of a local car-share scheme and their dependence on inefficient private vehicles for transportation. However, this should be counteracted

Table 6.7 Sensitivity analysis performed on a subsample of households with exactly 1 Earth and 2 Earths (including original results from Table **6.5**)

1 Earth (1)															
ID	**EOD 20**	**CL**	**GL**	**FL**	**FG**	**BL**	**cF**	**F**	**H**	**M**	**G**	**S**	**EF**	**CF**	**%**
SE39	13 Dec	0.4	0	0.4	0	0.1	0.8	0.4	0.7	0.1	0	0.5	1.8	2.2	42
QA09	22 Dec	0.5	0	0.2	0	0.1	0.9	0.5	0.4	0.4	0	0.4	1.7	2.6	51
SE39	21 Jan*	0.3	0	0.4	0	0.1	0.7	0.3	0.7	0.1	0	0.4	1.6	2	43
QA09	23 mar*	0.3	0	0.2	0	0.1	0.7	0.3	0.4	0.3	0	0.3	1.4	2.1	53
1 Earth (3)															
ID	**EOD 20**	**CL**	**GL**	**FL**	**FG**	**BL**	**cF**	**F**	**H**	**M**	**G**	**S**	**EF**	**CF**	**%**
QA09	28 Dec	0.5	0	0.3	0	0.1	0.8	0.5	0.4	0.3	0	0.5	1.7	2.4	49
2 Earths (1)															
ID	**EOD 20**	**CL**	**GL**	**FL**	**FG**	**BL**	**cF**	**F**	**H**	**M**	**G**	**S**	**EF**	**CF**	**%**
SE10	26 Jun	0.8	0.1	0.5	0.1	0.2	1.9	0.8	0.6	0.5	0.9	0.7	3.5	5.4	53
MO02	30 Jun	0.6	0.1	0.5	0	0.2	2	0.3	0.5	0.2	1.7	0.7	3.4	5.9	59
SF01	28 Jun	0.6	0	0.6	0	0.2	2	0.4	0.7	0.5	1.1	0.7	3.4	5.8	58
SE10	21 Aug	0.5	0	0.5	0	0.1	1.4	0.4	0.6	0.3	0.9	0.5	2.7	4.2	54
MO02	16 Jul	0.5	0.1	0.5	0	0.2	1.9	0.2	0.4	0.2	1.7	0.6	3.1	5.5	60
SF01	05 Jul	0.6	0	0.6	0	0.2	1.9	0.4	0.7	0.5	1.1	0.6	3.3	5.6	58
2 Earths (2)															
ID	**EOD 20**	**CL**	**GL**	**FL**	**FG**	**BL**	**cF**	**F**	**H**	**M**	**G**	**S**	**EF**	**CF**	**%**
SE10	15 Nov	0.5	0	0.3	0.1	0.1	0.9	0.6	0.6	0.3	0	0.5	1.9	2.5	45
MO02	23 Feb*	0.4	0	0.2	0	0.1	0.7	0.5	0.4	0.2	0	0.4	1.5	2.1	48
SF01	09 Nov	0.3	0	0.4	0	0.1	1.1	0.3	0.7	0.5	0	0.5	2	3.1	54
2 Earths (3)															
ID	**EOD 20**	**CL**	**GL**	**FL**	**FG**	**BL**	**cF**	**F**	**H**	**M**	**G**	**S**	**EF**	**CF**	**%**
SE10	22 Jul	0.7	0	0.5	0.1	0.1	1.6	0.7	0.6	0.3	0.9	0.6	3	4.6	52
MO02	08 Oct	0.6	0	0.3	0	0.1	1.1	0.5	0.5	0.2	0.6	0.5	2.2	3.3	51
SF01	27 Jul	0.5	0	0.5	0	0.2	1.7	0.3	0.7	0.4	0.9	0.7	3	4.9	56

[a]2021

Fig. 6.4 Photos showing recycling (plastic, cardboard, and cans) and reuse of tires. (Alexander Skutch Biological Corridor, photos by M. Thornbush, March 2020)

by the fact that they are producing biomass by cropping (coffee, sugarcane, grass, etc.), which fixes carbon. It should also be noted that this contribution that these producing households make to production stands in contrast to urban households that are not engaging in urban agriculture.

6.5 Conclusions

This chapter has revisited the general findings from the case study in the corridor. It has drawn from these results to present tailored suggestions to reduce the Ecological Footprint, especially as regards the crop land component and food consumption category – although not exclusively. Notably, it is important to consume food that is locally grown and encourage organic farming in the study area. As observed in the previous chapter (see Chap. 5), there is already an interest among these small-scale farmers to use organic fertilisers and reduce the amount or frequency of pesticide application. Organisations such as CoopeAgri can encourage this transition by either reducing taxes and/or subsidising farmers wanting to make the switch. It is also important to educate these farmers on organic methods as well as waste management. Charging for garbage, but not recycling, combined with farm-based composting works well. However, there are still misconceptions about burying garbage and/or burning it (especially paper) that need to be addressed.

References

Delahaye J (2019) Costa Rica named UN Champion of the Earth in fight against climate change. Mirror (24 September 2019). https://www.mirror.co.uk/travel/central-south-america/costa-rica-named-un-champion-20166387

Galli A, Wiedmann T, Ercin E, Knoblauch D, Ewing B, Giljum S (2012) Integrating ecological, carbon and water footprint into a "footprint family" of indicators: definition and role in tracking human pressure on the planet. Ecol Indic 16:100–112. https://doi.org/10.1016/j.ecolind.2011.06.017

Global Footprint Network (2020) National Footprint Accounts. https://data.footprintnetwork.org/#/

Holmberg J, Lundqvist U, Robèrt K-H, Wackernagel M (1999) The ecological footprint from a systems perspective of sustainability. Int J Sustain Dev World Ecol 6(1):17–33. https://doi.org/10.1080/13504509.1999.9728469

Kitzes J, Moran D, Galli A, Wada Y, Wackernagel M (2009) Interpretation and application of the ecological footprint: a reply to Fiala (2008). Ecol Econ 68(4):929–930. https://doi.org/10.1016/j.ecolecon.2008.11.001

Levett R (1998) Footprinting: a great step forward, but tread carefully — a response to Mathis Wackernagel. Local Environ 3(1):67–74. https://doi.org/10.1080/13549839808725545

Lin D, Hanscom L, Murthy A, Galli A, Evans M, Neill E, Mancini MS, Martindill J, Medouar F-Z, Huang S, Wackernagel M (2018) Ecological footprint accounting for countries: updates and results of the National Footprint Accounts, 2012–2018. Resources 7(3):58. https://doi.org/10.3390/resources7030058

Wackernagel M (1998) The ecological footprint of Santiago de Chile. Local Environ 3(1):7–25. https://doi.org/10.1080/13549839808725541

Wiedmann T, Minx J (2008) A Definition of 'Carbon Footprint.' In: Ecological economics research trends. Nova Science Publishers. https://www.novapublishers.com/catalog/product_info.php?products_id=5999

Wright LA, Kemp S, Williams I (2011) 'Carbon footprinting': towards a universally accepted definition. Carbon Manage 2(1):61–72. https://doi.org/10.4155/cmt.10.39

Chapter 7
Conclusion

Abstract In this final chapter, the chief contributions of this brief are presented and function to further justify the research. A deliberation of the research limitations also appears here, before suggesting ideas for future research. Finally, there are some concluding remarks for wrapping up this brief.

Keywords Case study · Alexander Skutch Biological Corridor · Costa Rica

7.1 Contribution

This study has considered various criticisms of the Ecological Footprint and biocapacity accounting with a focus on the Footprint Calculator. In other words, it has presented the issues from a critical approach. Such an approach has involved scrutinising the Footprint Calculator as the instrument of choice in the field surveys as well as the National Footprint Accounts, on which they draw. These tools are not without their problems, as for instance noted in the carbon component of the National Footprint Accounts that attributes zero to biocapacity. The latter needs to be incorporated into the spreadsheet (e.g., Country_Trends for Costa Rica – country_code = 48). This issue is also reflected in the Footprint Calculator – that does not measure production (biocapacity) and only the Ecological Footprint (consumption). Such criticisms have been drawn from a variety of sources, including published literature and other unpublished sources (e.g., Jira – Issue Tracker in Chap. 1, see Sect. 1.4.1). One of the chief criticisms, for instance, has been the focus on consumerism in the developed world – that has augmented the carbon Footprint in particular. In contrast, the brief's study area is in a developing country, where participants were chosen from peasant farmers '*campesinos*' who are struggling with poverty. In this way, the issue of poverty has entered the research, making the instrument relevant not only to environmental issues but also socioeconomic ones. However, the Ecological Footprint is simply deficient in this area as an environmental composite indicator (Strezov et al. 2017). There is potential to

M. J. Thornbush, *The Ecological Footprint as a Sustainability Metric*,
SpringerBriefs in Environmental Science,
https://doi.org/10.1007/978-3-030-62666-2_7

incorporate more of the human impacts on the environment, nevertheless, by considering the environmental quality of managed ecosystems (e.g., Erb 2004).

More specifically, the case study in the corridor has helped to ascertain the issues currently confronting farmers in the area. For example, many expressed the need for further education in recycling and waste management – to supplement what the Blue Flag Ecological Program is already doing in the area to raise awareness of environmental issues (refer to Sect. 5.4.5). Education is also needed in the academic sector to improve the Ecological Footprint and biocapacity accounting, especially to assist in understanding its limitations (this has been recently tackled by the Footprint Data Foundation by way of its recently formed SAC). This could also form part of future research and outreach associated with such endeavours. The current study has already solicited issues from farmers through commentary provided at the end of the interviews. It is suggested here that the recommendations in this study are presented to CoopeAgri to give back to the community by helping them to fulfil their requests for further education in waste management (relevant to the Blue Flag Ecological Program) and to provide more financial support for organic farming and reduced taxes (e.g., down from 8%) in attempts to bring them out of poverty. Moreover, it is possible to give back to the participating communities (corridor towns) in the case study through a presentation that relays the study findings while contributing to outreach and education.

Although an interesting exercise, the investigation has not been able to change the survey and only make recommendations (in Sect. 7.4) on what needs to be altered. To make the necessary changes, the survey would no longer be comparable to existing results (cf. Jóhannesson et al. 2020), and would prohibit any cross-temporal comparison based on past results. As it is, cross-temporal comparisons based on the National Footprint Accounts are necessarily edition-specific because of the constraints emplaced by changing equivalence factors. It is, therefore, problematic to instigate the changes needed for the Footprint Calculator to become less development biased. This is also a problem that exists in the National Footprint Accounts – on which the Footprint Calculator is calibrated for national scores – that cannot be corrected, but needs to be understood as part of the applicability of the instrument and its interpretative error. For instance, developed nations are usually the ones with better quality scores or the ability to improve their data through available government agencies, as indicated on the quality scores website (www.footprintnetwork.org/data-quality-scores) and addressed in Chap. 1 (see Sect. 1.4.4.1).

7.2 Limitations

It is noteworthy that as the selected research instrument, the Footprint Calculator is not perfectly tailored to the corridor, and limitations of deploying this tool to measure the Ecological Footprint of farming households in the Alexander Skutch Biological Corridor is an important consideration. As detailed in Chap. 1 (see Sect. 1.4.1), the questions are not equally distributed among the three categories (of food, housing, and transportation), with food (and to some extent transportation) not receiving enough attention. Some other questions that could be asked in the food category, for example, could address organic food production and consumption, quantities of food produced and consumed, types of processed foods consumed, and how exactly food is being acquired for consumption if not produced by the consumer. It should also be noted that the Footprint Calculator, in its current form, does not account for any geographical or cultural anomalies in the data – specific to the study area and perhaps Costa Rica at large, as for instance that fans are used but not heaters or air conditioners. Furthermore, poverty is a socioeconomic issue that may not manifest in the outputs but, nevertheless, is an underlying issue affecting the responses – as to why people do not fly every year, have inefficient vehicles and houses, eat meat and fish once per week and consume more beans, rice, and eggs (although this is also cultural), and are not purchasing much clothing and consuming many items (like newspapers, etc.), although this could be attributable to environmental ethics in the corridor.

This study has been limited in ways already noted by criticisms posed towards the National Footprint Accounts and Footprint Calculator (see Chap. 1, Sect. 1.4.1). However, there are also limitations posed by the case study. For one, as a case study it does not reflect all places in the world and can only truly represent itself – as part of a rural environment in a developing country, where peasant farmers or '*campesinos*' were asked to complete the survey-interviews. It is also questionable how generalisable the findings can be to other tropical countries (and other developing countries), as this was not an explicit study ambition. Participants in the study area were sampled according to where they resided in the corridor, as along the main roads in towns. Snowballing was also deployed, to some extent at the sampling locations, and this has its own issues – as for example people will tend to recommend people they know, who could be more like them and, therefore, reduce diversity in the sampled population. On the other hand, neighbours are also not always aware of who their neighbours are and their information may be misleading – or they could not easily answer questions that compare them to other neighbours contained in the Footprint Calculator survey (e.g., Question 10). This conveys the urban-centricity of the instrument because it assumes that neighbours are within a close proximity to each other, which is not normally the case in a rural setting.

Importantly, there were limitations imposed by time and funding, so that it was not possible to sample more corridor households, as in the towns of Santa Marta, San Ignacio, and Montecarlo (see Table 6.4). In hindsight, it would have been more prudent to reduce the number of samples in Santa Elena and sample more of the households in these towns (Santa Marta, San Ignacio, Montecarlo). Trinidad is a '*barrio*' or neighbourhood in Santa Elena, so was treated as part of this town in the case analysis. To reduce error between researchers or surveyors doing the fieldwork, the number of people who executed the survey-interviews were kept to a minimum so that the same people could, as consistently as possible, deliver the survey-interviews. However, this inadvertently limited the number of people that could be included in the study – both as surveyors and participants – and impacted the sample size.

7.3 Future Research

More research is needed to decipher the impact that income has on answering questions in the Footprint Calculator survey, as poverty has been shown to affect how questions are approached and answered. Similarly, more work is needed to study the impact of industrial development (as it affects income) on the Footprint Calculator. Moreover, it would also be interesting to test the survey on urbanites in countries with different levels of development to see whether city-dwellers are any closer to the 'target audience' than those participants in the current study, who were rural-dwellers (townspeople) in a developing country. This study has already contributed to doing this by examining rural communities. Importantly, although some authors (e.g., Strezov et al. 2017) have classified the Ecological Footprint as wholly in the environmental dimension of sustainable development (see their Table 2 and Fig. 2, p 248), the social and economic dimensions are also indirectly represented through the Footprint Calculator, as responses are influenced by the socioeconomic situation (including poverty) as well as cultural aspects of participants. Language is not the only factor that restrains participation, as poverty could impose illiteracy – including computer illiteracy – and limited access to the Internet could also impact who uses the Footprint Calculator online. Lastly, this study did not execute ANOVA, but this statistical analysis could be incorporated into future studies deploying the Footprint Calculator through its components and categories. More studies are needed to examine the Ecological Footprint (and biocapacity) in Costa Rica to add to the published literature (e.g., Adams and Ghaly 2006 for agroindustries in its coffee industry; Kissinger and Rees 2010 regarding trade; van Vuuren and Smeets 2000). In particular, it would be interesting to continue to develop more improvement scenarios, as presented in Chap. 6 (Sect. 6.3.1).

7.4 Recommendations

There are some implications of this study. For one, it is evident based on the case study that more government intervention is needed to support rural farmers in the corridor, as poverty is clearly evident in this study area. The major buyer of produce from these rural farmers is CoopeAgri (and VolCafe in Montecarlo, as well as other organisations listed in Fig. 5.7). There is an impetus for government to work with the cooperatives more closely and oversee their business with the local farmers. For example, one of the farmers (QA19) suggested that there are existing funds that farmers cannot access, even if they may quality for them, because forms need to be completed and the cooperative is not helping them with administrative outreach.

As regards the Ecological Footprint, their high scores on the crop land component and food consumption category, in particular, convey their reliance on the land for food production and consumption. On the other hand, grazing land and fishing grounds were generally not high scores among the participants. They are similar to rural-dwellers anywhere else in the world in that they depend on private vehicles that are inefficient but are, nevertheless, used to get around town and to work. It would help to have them reside on their farms in order to reduce the amount of work-related travel time associated with getting to work in the farmlands. Household heads were not accustomed to taking public transport, however, although some of their household occupants (mainly wives) did use it when necessary.

Stemming from the follow-up interviews, it was evident that produce yields did not necessarily reflect hectarage or income. This indicates that there is varied intensity of land use in the study area, which can be proxied using mass over area data (see Chap. 5). It was also discovered during the course of the fieldwork that CoopeAgri keeps tracks of yields produced in the corridor (in a database) and this could be deployed somehow to supplement information on local biocapacity based on what is locally produced for the region. This would help to inform the biocapacity portion of the accounts that is currently excluded (and not addressed) by the Footprint Calculator. Another highlight here is the necessity to improve the livelihoods of these local producers (e.g., through efforts towards poverty alleviation) and improve soil health and fertility as well as produce quality (e.g. support organic production).

7.4.1 Applying the Footprint Calculator

It is necessary to test the Footprint Calculator in developing regions in order to overcome any existing shortcomings. The chief concern in altering the Footprint Calculator is that the actual survey cannot be changed in order for it to maintain its comparability across nations and years (in the case of repeated studies) – a criticism also cf. Jóhannesson et al. (2020). It should be noted, however, that changes have already been made to the National Footprint Accounts since their inception based on published criticisms and suggestions for improvement – using equivalent factors is an example of this. For the accounts, this means that the edition needs to be specified in the published literature relaying findings that are specific to an annual edition (the 2019 Edition was deployed in the current study). Arguably, however, it is not enough to 'calibrate' using the National Footprint Accounts if the survey questions are inappropriate, biased, or blind in some regard that impacts the results. A conundrum, therefore, exists in that it is necessary to 'acculturate' the Footprint Calculator survey (and change it in order for it to be improved) but, on the other hand, this could restrict and nullify cross-temporal and/or -spatial comparisons. What can be done to overcome this perhaps is to encourage people to complete the Footprint Calculator with every new edition of the National Footprint Accounts as a way to gauge cross-temporal (behaviour) change.

In the current application of the case study conducted in the corridor, for instance, not all questions were relevant (and they were assumed – and a consistent default used, as evident in Appendix 1.1 – and not asked in order to reduce the survey time). These included questions pertaining to carpooling (when no car-share scheme was formally in place) or the amount of renewable energy used (with the national constant of 95% deployed). Moreover, participants did not fly on a regular basis – each year, so that the question asking how often participants fly each year quickly became irrelevant. The same was true for the use of trains, where there is no train service in the area. In addition, most of their vehicles are inefficient, so this was kept as the default option because most people could not provide this information. All houses sampled in the corridor have electricity and are made from either brick or concrete or, otherwise, wood, so this was likewise reflected in the questions asked. Similarly, in all cases, their houses were freestanding and had running water. Finally, when performing fieldwork based on such survey-interviews, it is important to keep it short – and even 30 min was too long for most people. Therefore, it is advised that the entire process be kept brief, up to 20 min. This means ascertaining defaults for the sample and dropping some questions wherever and if possible. From experience, people responded well when told that the entire survey-interview process would take no longer than 20 min of their time.

7.5 Concluding Remarks

The use of a detailed case study in this research greatly helped to contextualise and test the application of the National Footprint Accounts through the Footprint Calculator. More could be done to acculturate the survey, more than translating it to Spanish and other languages. Studies such as these help to calibrate the instrument to socioeconomic as well as cultural factors that will undoubtedly impact responses and performance. Such regional-scale studies also inform national results and offer a means of differentiating the top-down approach through bottom-up testing and verification. It is necessary that more work is executed, ultimately, to provide confidence levels or limits – such as a confidence interval of 95%, as in this study – as quality indicators used in conjunction to information on estimation procedures (sampling size, measurement protocols, quality control procedures, and so on, cf. Birigazzi et al. 2019 see Jóhannesson et al. 2020; Kitzes and Wackernagel 2009) to the Ecological Footprint and biocapacity accounts in order to make explicit the error associated with the data and its application. Lastly, this brief has revealed socioeconomic and cultural aspects of the study sample through responses provided in the case. The case study has been instrumental to not only identify further research and outreach activities still needed, but also to convey some existing limitations evident even in the Spanish version of the Footprint Calculator as the instrument deployed in the field investigation. While it is important to consider the issues with the National Footprint Accounts as an input into the Footprint Calculator, calibrating the instrument (as a global Footprint calculator) to measuring different cultural perspectives is difficult. The main reason for this is that changing questions for specific versions (e.g., Spanish) would create a disjuncture between other versions, making any comparison inappropriate. However, it is possible to tailor it to some extent through certain omissions and defaults understood to represent general responses based on preliminary testing. In this case, our field guide (Alex Fonseca) was chosen as the first study participant in the training process. This included getting the surveyors ready for the field surveys, but also helped to tailor the survey instrument to the study participants (audience). One of the key recommendations was to keep the process as brief as possible in order to retain participant attention and cooperation.

References

Adams MA, Ghaly AE (2006) An integral framework for sustainability assessment in agro-industries: application to the Costa Rican coffee industry. Int J Sustain Dev World Ecol 13 (2):83–102. https://doi.org/10.1080/13504500609469664

Birigazzi L, Gregoire TG, Finegold Y, Golec RDC, Sandker M, Donegan E, Gamarra J (2019) Data quality reporting: good practice for transparent estimates from forest and land cover surveys. Environ Sci Pol 96:85–94. https://doi.org/10.1016/j.envsci.2019.02.009

Erb K (2004) Actual land demand of Austria 1926 – 2000. A variation on ecological footprint assessments. Land Use Policy 21(3):247–259. https://doi.org/10.1016/j.landusepol.2003.10.010

Jóhannesson SE, Heinonen J, Davíðsdóttir B (2020) Data accuracy in ecological footprint's carbon footprint. Ecol Indic 111:105983:9. https://doi.org/10.1016/j.ecolind.2019.105983

Kissinger M, Rees WE (2010) Exporting natural capital: the foreign eco-footprint on Costa Rica and implications for sustainability. Environ Dev Sustain 12(4):547–560. https://doi.org/10.1007/s10668-009-9210-7

Kitzes J, Wackernagel M (2009) Answers to common questions in ecological footprint accounting. Ecol Indic 9(4):812–817. https://doi.org/10.1016/j.ecolind.2008.09.014

Strezov V, Evans A, Evans TJ (2017) Assessment of the economic, social and environmental dimensions of the indicators for sustainable development. Sust Dev 25:242–253. https://doi.org/10.1002/sd.1649

Van Vuuren DP, Smeets EMW (2000) Ecological footprints of Benin, Bhutan, Costa Rica and the Netherlands. Ecol Economics 34(1):115–130. https://doi.org/10.1016/S0921-8009(00)00155-5

Index

A
Acculturate, 101, 102
Aggregation, 10–12, 14–16, 56
Alexander skutch biological corridor ('corridor'), v, 2, 15, 48, 51, 54–57, 62, 68, 81, 82, 88, 95, 98
Amacobas, 77, 79, 89
Arable farming, 74, 82
Auction, 77

B
Beans, 89, 99
Biological capacity (BC)\biocapacity, v, 1–3, 5–8, 10, 13–15, 18–20, 31–34, 36–38, 41–48, 51–56, 68, 69, 74, 81, 82, 90, 92, 97, 98, 100–102
Biological productivity\bioproductivity, v, 2, 5, 12, 32, 45, 54, 62, 69
Built-up land, v, 2, 5, 7, 14, 15, 35, 45, 54, 64, 69, 72, 90, 92

C
Cacao (cocoa), 74, 76, 83
Carbon
  carbon dioxide(CO_2), 5
  carbon footprint (CF), 5, 13, 32, 36, 44, 46, 54, 68, 69, 71, 82, 86, 87, 89, 90, 92, 93, 97
Carpooling/-sharing, 60, 71, 101
Carrying capacity, 2, 31, 35, 38, 41–45, 48
Case study, v, 2, 16, 18, 20, 48, 51–63, 67–80, 82, 96, 98–102
Cattle (cows), 74, 76, 82, 83
Chickens, 21, 76
City-dwellers, 100
Climatic bias, 18
Coffee, 57, 74–77, 83, 95, 100
Communities, 5, 33, 55, 58, 62, 78, 79, 87, 93, 98, 100
Components, 3–5, 9, 10, 12, 18, 19, 32, 35–37, 45, 46, 53, 54, 62, 64, 68, 71, 72, 74, 82, 85, 87, 89, 90, 92, 93, 96, 97, 100
Consent, 58, 59, 61, 63
Conservation, 34, 79
Constants, 1, 54, 102
Consumption
  categories, v, 14, 55, 56, 60, 61, 68, 69, 72, 73, 83, 85, 90, 96, 98–100
  consumers, 99
  consumes, 46, 55, 62, 96, 99
Consumption Land Use Matrix (CLUM), 61
Cooperatives
  CoopeAgri, 100
  CoopeCedral, 77
Cover crops, 78
Critical approach
  criticisms, 10, 56
Crop land, v, 2, 5, 7, 15, 32, 35, 46, 53, 64, 69, 71, 72, 82, 90, 92, 93, 96, 100
Crops, 7, 45, 46, 57, 61, 65, 74–78, 82–84
Cultural aspects, 14, 36, 48, 100, 102

M. J. Thornbush, *The Ecological Footprint as a Sustainability Metric*,
SpringerBriefs in Environmental Science,
https://doi.org/10.1007/978-3-030-62666-2

D
Data quality, 7, 10, 11, 14–18
Development
  developed countries, 17, 46
  developing countries, 16, 17, 47, 99, 100
  sustainable development, 2, 100
  United Nations Sustainable Development Goals, 3, 52
Double-counting, 11, 13

E
Ecological debt/deficit, v, 4, 33, 42, 46, 54
Ecological footprint (EF), v, xi, 1–13, 15, 16, 18–20, 31–38, 41–48, 51–56, 60, 62, 64, 68–74, 80–82, 85–92, 94, 96–98, 100, 102
  ecological footprint of consumption (EF_C) consumption Footprint, 2, 3, 5, 8, 32, 36, 41, 42, 45, 51, 53, 54, 56, 70, 90
  ecological footprint of exports (EF_E), 7
  ecological footprint of imports (EF_I), 8
  ecological footprint of production (EF_P) production Footprint, 3, 5, 32, 36, 46, 51, 82, 97
Ecological or biocapacity deficit/reserve, v, 32–34, 48, 55
Ecological overshoot, 4, 13, 20, 32–34, 42, 46–48, 51, 54, 68, 82
Economic dimensions, 100
Eggs, 21, 76, 89, 93, 99
Energy-centrism, 11, 13–15
Energy use, 13, 14, 16, 17, 101
Environmental dimensions, v, 2, 36, 100
Equivalence factor (EQF), 6–9, 14, 32, 33, 98

F
Fair Trade, 61, 65, 77, 79, 83
False concreteness, 11, 12, 15
Fertilisers, 61, 65, 78, 83, 96
Fishing grounds, v, 2, 5, 7, 15, 32, 35, 45, 53, 64, 69, 71, 72, 82, 90, 92, 100
Flying\flights, 60
Foods, v, 2, 4, 6, 14, 16, 17, 21, 42, 45, 53, 55, 56, 60, 61, 64, 68, 69, 71–73, 81, 83, 87–90, 92, 93, 96, 98–100
Footprint Calculator (Ecological Footprint Calculator), v, 2, 10, 11, 14–18, 20–23, 33, 36, 51, 54–56, 58–63, 68, 69, 71, 73, 74, 81, 87, 93, 97–102
Forest lands, v, 2, 5, 7, 9, 15, 32, 35, 45, 53, 64, 72, 90, 92
Fruits, 74, 76

G
Garbage, 17, 78, 93, 96
Global hectare (gha), 6, 69
Grass (for cattle/cows), 74, 76, 78, 83, 95
Grazing lands, v, 2, 5, 7, 15, 32, 35, 45, 53, 64, 69, 71, 72, 82, 89, 90, 92, 100

H
Health, 4, 35, 83, 92, 93, 101
Households
  heads, 61
Housing, v, 14, 16–18, 21, 55, 56, 60, 61, 64, 68, 69, 72, 73, 87, 90, 92, 98

I
Incomes, 54, 60, 63, 69–71, 76, 83–85, 100, 101
Inefficient vehicles, 99
Infrastructure, 7, 45
Intertemporal yield factor (IYF), 6, 8, 9

L
Land uses, 6–8, 11–15, 35, 45, 61, 71, 101
Lychees, 74, 76, 83

M
Manure, 78, 83
Markets, 34, 53, 77
Meals, 89
Mixed crops, 83
Monocrops/-cropping/Monocultures, 74, 83
Montecarlo, 57, 69, 70, 77, 85, 86, 99, 100

N
Natural capital, v, vi, 4, 5, 7, 31, 34, 36, 41, 42, 45, 46, 54, 82
Neighbourhoods, 57, 99
Neighbours*barrios*, 17, 22, 57, 77, 99

O
Organic farming, 61, 65, 77, 83, 96, 98
Organics, 16, 58, 77, 78, 83, 89, 96, 99, 101

P
Participants, 16, 17, 55, 57–60, 62, 63, 68, 69, 76–80, 82–84, 89, 90, 97, 99, 100, 102, 103

Pastoral farming, 74, 82
Peasant farmers ('*Campesinos*'), 15, 48, 57–60, 80, 89, 97, 99
Peppers, 74, 76, 83
Pesticides, 61, 65, 78, 83, 93, 96
Poverty, 17, 90, 97–101
Production
  produces, 76, 77
  products, 8, 45, 46, 77, 82
Productivity
  primary product, 5, 8, 10, 42, 45
  secondary product, 35

Q
Quality
  assurance (QA), 18
  scores, 10, 18, 98
Quizarrá, 57, 70, 77, 85, 89, 90

R
Recycling, 71, 89, 92, 93, 95, 96, 98
Renewable energy\renewables, 14, 16, 17, 21, 22, 52, 60, 88, 89, 102
Reuse, 71, 93, 95
Rice, 89, 99
Roads, 87, 88, 99
Roots, 76
Roya, 78, 83
Rural-dwellers, 100

S
Sampling
  field/study sites, 56–58
  sample locations, 69, 70, 99
  snowball sampling\snowballing, 57, 99
  survey-interviews, 99
  surveyors, 57–59, 74, 99, 103
San Francisco, 57, 69, 70, 77, 85, 86
San Ignacio, 57, 69, 70, 85, 99
San Isidro del General, 87
Santa Elena, 57, 69, 70, 74, 85, 90, 99
Santa Marta, 57, 69, 70, 74, 77, 85, 86, 99
Scale (spatial/temporal), v, 3, 10–12, 14–16, 32–36, 42, 55, 56, 60, 68
Schools, 35, 57, 77
Social dimension, 100
Socioeconomics, 2–4, 31, 32, 36–38, 48, 54, 92, 97, 99, 100, 102
Soil
  fertility, 33, 101
  health, 83, 101
Sugarcane, 74, 76, 83, 95
Sustainability
  environmental sustainability, 37, 51
Systems (theory), 1, 3–5, 13, 31, 32, 34–37, 56

T
Towns or ('*Pueblos*'), 57
Trade, 2, 5, 6, 10, 34, 37, 38, 100
Transportation
  public transportation, 17, 22, 89
Trees (fruit), 45, 46, 77–79
Trinidad, 57, 99

U
Urbanisation/urbanism, 35
Urbanites, 17, 18, 100
Utility
  theory, 13

V
Vegan, 92, 93
Vegetables, 74, 76, 77
Vegetarian, 88, 92, 93
Volcafe, 77, 100

W
Waste management, 93, 96, 98

Y
Yield factors, 8, 9, 14, 15, 33
Yields, 8, 9, 14, 75, 76, 83, 84, 101